BOILER TEST
CALCULATIONS

Boiler Test Calculations

JOHN SENIOR
C.Eng. M.I.Mech.E.

Edward Arnold
A division of Hodder & Stoughton
LONDON   NEW YORK   MELBOURNE   AUCKLAND

© 1989 John Senior

First published in Great Britain 1989

British Library Cataloguing in Publication Data

Senior, John
   Boiler test calculations
   1. Industrial boilers. Thermal properties.
   Testing. Calculations
   I. Title
   621.1'84

   ISBN 0-85264-301-2

Typeset in 10/12pt Times by Unicus Graphics Ltd, Horsham.
Printed and bound in Great Britain for Edward Arnold, the
educational, academic and medical publishing division of Hodder
and Stoughton Limited, 41 Bedford Square, London WC1B 3DQ
by Richard Clay Limited, Bungay, Suffolk.

Contents

CONTENTS

Figures

Tables

Preface

The thermal testing of boilers is of crucial importance to those who manufacture or install them, and also — in particular — to those who own or operate them. This concern stems from the quest for the highest possible thermal efficiency, in the one case as a valuable selling feature and in the other as a means of reducing energy costs.

"Testing" may mean much or little, but whether it is extremely comprehensive or merely the determination of a few principal characteristics, for the results to receive general acceptance they should be based upon one or other of the standard test procedures specified by organizations such as BSI or ASME. For the sake of brevity, standard test codes usually present much of their guidance and formulae without explanation, and the user may on occasion be uncertain why a particular approach has been adopted. This book has been written as a companion to the boiler test codes used in Britain and America: the development of fundamental and simplified test formulae is explained in detail and attention is drawn also to points that are controversial or subject to alternative treatment. As an aid to non-specialized readers, some general information on fuel and instrumentation has also been included. For simplicity, it is assumed throughout the book that the heat transfer medium under consideration is water or steam/water, but all relevant formulae can be assumed to apply equally to other heat transfer fluids, provided suitable allowance is made for their different physical characteristics.

The symbols and units used in the text are listed in Table A.1 in Appendix 1. For simplicity, and as an aid to anyone wishing to use them in computer programs, none of the symbols representing physical quantities include Greek letters, lower-case letters, subscripts, superscripts or indices. Any numerals or letters used in parenthesis after symbols to distinguish different categories, for example, MH2(1), are explained at the point of use.

Acknowledgements

Grateful acknowledgement is made to the British National Committee of the World Energy Conference for permission to use data from their publication *Technical Data on Fuel*, and to British Coal and the College of Fuel Technology for permission to reproduce data on the classification and sizes of coals. Extracts from British Standards are reproduced by kind permission of the British Standards Institution: complete copies can be obtained from BSI at Linford Wood, Milton Keynes MK14 6LE.

1. Types of test

1.1 Aims of test

Essentially, the thermal testing of a boiler involves the determination of its thermal efficiency, that is, the proportion or percentage of the heat input to the boiler which is available as useful output, and also the thermal output or rating obtained from it at the time the efficiency was measured. To state the efficiency alone is not very informative — rather like giving the fuel consumption of a car without declaring the speed at which that fuel consumption was obtained. Determination of the thermal input and various heat losses may also be required, perhaps as factors necessary in the calculation of the efficiency and output, or to provide information concerning specific aspects of the boiler's performance.

1.2 Full heat-account tests

Several options are available when selecting a method for testing a boiler. Ideally there is much to be said in favour of taking measurements to establish heat input, heat output, and all losses. Obviously this procedure is self-checking because the input should be equal to the sum of the output and the losses, and any discrepancy obliges the test staff to seek for the reason. Also, with every loss detailed, it is immediately evident whether any one is abnormally high, and by making appropriate operational or other changes to reduce it, the thermal efficiency can be raised.

The tabulated data obtained from such a test are usually referred to as a "full heat balance" or "full heat account". Ideal it may be, but a test as comprehensive as this is rarely carried out. More often the closest approximation to such a test is one in which one or more losses — usually radiation loss is one of them — is not directly determined but is found "by difference", that is, as the balance of the heat account. Obviously this destroys the self-checking aspect of the heat account because any errors are hidden in those losses found by difference. The reason for adopting this procedure is simply one of expense; any reasonably comprehensive boiler test is expensive, and it may also be inconvenient when carried out on-site, so there is a natural desire on the part of test staff to by-pass any arduous and time-consuming procedures. However, even in this modified procedure, the test staff are still committed to the direct determination of the input, output, and — wherever possible — most of the losses associated with the boiler.

1

1.3   Indirect tests

A more fundamental variation in test procedure is employed in an "indirect" test. Using this method, either the input or the output is obtained not from direct measurements of the physical quantities but is instead derived indirectly by calculation, using information relating to the other heat flows. It is in fact another variation of the "by difference" technique referred to previously.

A typical example of how this method may be applied is provided by a test on a boiler generating saturated steam. In a test house the output from such a boiler, if it were not too large, might be measured by condensing all of the steam and calculating the heat content from the rise in temperature of a measured quantity of condenser-circulating water. This would be impractical if the boiler were too large or if the test were carried out on site. Yet, obtaining a representative sample of saturated steam is also impossible because the moisture content tends to slide along the walls of the steam pipe and by-pass any sampling probe which is used. Such practical difficulties would lead test staff to adopt the indirect method, and derive the boiler output from the input and losses, these being derived in turn from measurements of physical quantities.

Similarly, the reverse situation could apply, and in the case of a boiler fired by solid fuel it might be judged impractical to measure accurately the quantity of fuel burnt (and hence the heat input) over the test period, whereas if the boiler in question was producing hot water or superheated steam, the heat output could be measured to an acceptable accuracy. Again, by use of the data obtained from measurements, the unknown quantity — in this case the heat input — could be derived.

1.4   Direct versus indirect tests

At times, the relative merits of the direct and indirect methods are discussed with considerable partisan fervour, the advocates of indirect testing usually claiming that, especially in the case of on-site testing, measurement of input or output is impracticable. It is also claimed that in any case, with a modern boiler having an efficiency of 80% or better, it is intrinsically more accurate to measure the losses rather than the input or output because the sum of the losses probably amounts to less than 20%, and so any departure from precise measurements of the physical quantities involved — temperature, pressure, flow, and so on — will have a correspondingly smaller effect on the final result.

Against this, it has to be recognized that it is not only direct testing procedures which may be difficult to follow on-site. Boilers are often installed without any regard for test requirements and, for example, short or tortuous flue-gas ducting may render it extremely difficult to obtain reliable flue-gas samples or temperatures for determination of dry flue-gas losses. Again, even in an indirect test carried out with great care, it will be highly unlikely that all possible losses will be established from measurements. Radiation losses are often assumed from recommended figures, and a number of minor losses such as those due to moisture in the combustion air, hydrocarbons in the flue gas, sensible heat in ash, carbon loss in fine dust and so on, may not be investigated. Small these individual losses may be, but cumulatively they may be significant. Sometimes "typical" figures are attributed to such losses, but if there is some malfunction or design fault, any single loss may be unexpectedly high, and the test should serve to establish this.

It is more realistic for test staff to be impartial concerning the relative merits of direct and indirect testing. At best, both should give the same result, and as a general rule it must always be more reassuring to obtain the maximum possible amount of information from a test, especially when this is of the contractual type, so that some measure of cross-checking is possible.

1.5 Type testing

A special category of thermal testing may be adopted in the case of boilers which are manufactured in ranges of standard or near-standard configuration and method of operation, which is commonly the case for small to medium sizes such as those of the sectional, welded steel and shell type, and also the smaller packaged or modular water-tube boilers. Boilers are selected at a number of points in the range of outputs available and undergo tests which are carried out according to a standardized procedure. These results are then accepted as being typical for the entire range. This procedure benefits the purchaser because it makes detailed test results available for examination; it benefits the manufacturer because it absolves him from testing every boiler sold; and it benefits both by reducing costs.

Type testing is essentially a matter of test policy rather than test procedure because a test carried out in accordance with any code may, by agreement, be accepted as a type test. However, as the concept is applied in particular to relatively small boilers, it is usual for type testing to be carried out at a test station in which special equipment and instrumentation is permanently installed in order to simplify test procedures. For example, special arrangements can be made for providing steady loads by means of cooling towers or other heat sinks for the units under test, or for accurately metering fuel and water.

At the time of writing, a British Standard for the thermal testing of heating boilers up to 600 kW $(2 \times 10^6$ Btu/h$)$ in rating is in course of preparation. The procedure is based upon the use of special test rigs suitable for testing this class of boiler.

1.6 Rudimentary forms of test

So far, the types of test described, whilst differing quite markedly in approach, all have the same basic aims — the determination to an acceptable level of accuracy of the thermal efficiency and output of a boiler. As explained in 5.4, it is desirable to obtain these over relatively short test periods. Whilst not the direct concern of this book, it is worth pointing out that it can nonetheless be quite informative to obtain average "in-use" efficiencies and outputs averaged over longer periods such as a day, a week, a month, a heating season, or a year. As averages, the figures obtained can be quite accurate and although not suitable for detailed analysis of boiler performance they can be of value for accounting purposes and will certainly direct the attention of the operators to broad trends in performance. Thus a comparison from one period to another of delivered fuel quantities with water or steam produced, or even with items of finished factory product, is a powerful indicator of thermal efficiency even if it is a broad average. Similarly, the widely applied (but unfortunately not usually well-maintained) instruments for flue-gas analysis and temperature measurements provide excellent indicators of boiler performance, either in the absolute sense or for simply indicating changes. If for no other purpose, flue-gas temperatures should be kept under scrutiny so as to obtain guidance on when boiler cleaning is necessary.

Simple "tests" are therefore hardly more than an extension of routine monitoring and, as stated, lie outside the scope of this book, but they are as important in their own way as the more detailed tests dealt with in test codes and should at least provide an excellent indication of when a more detailed test is necessary in order to establish the reason for any reduction in thermal efficiency.

2. Defining the boundary of a boiler unit

2.1 Definition of a boiler unit

It may seem somewhat pedantic to be concerned with the definition of such a familiar piece of equipment as a boiler, but the point is important. In earlier periods of engineering history there was little room for doubt — a boiler was a heat exchanger in which heat from fuel was transferred to a heat carrier or working medium which was water/steam. Boiling really did take place, as steam was evolved in the water. Later, when hot water alone came to be used for many heating purposes some purists objected to the word "boiler" being used to describe the "water heaters" or "hot water generators" which provided the hot water. Another type of boiler, in which water was pumped through much of the boiler structure in liquid form before flashing into steam came to be known as a "steam generator".

With the passage of time, "boiler" has come to be the general term favoured in ordinary usage for all these forms of heat exchanger, though, when special oils are used as the heat carrier, it is still fairly common to find "thermal fluid heaters" described as such.

Of much greater importance when it comes to assessing the thermal performance of a boiler is the matter of deciding where a boiler begins and ends. Does "boiler" include the combustion equipment, the oil heaters, the pumps, fans, and so on?

Essentially, at its most basic, a boiler is a heat exchanger, designed to transfer the heat supplied from some fuel or other energy source into the heat carrier. As far as possible therefore, other machinery, ductwork and pipework are not considered as integral parts. However, for theoretical or practical reasons or perhaps merely convenience or convention, other items do get included, and so it has come about that "unit" is often used as a general descriptive term in place of "boiler". The boundary of a "unit" under test is therefore likely to include a number of items other than heat exchanger components, and what these further items comprise will often be a matter for discussion and agreement, or disagreement, between the parties involved in the test.

Obviously when a unit is being commissioned one person who is vitally concerned is the manufacturer, who will wish it to work well, not only for contractual reasons but also from the promotional aspect, because the performance of his unit will be compared with that of others by prospective purchasers. He will naturally object if he feels that the overall performance will be reduced by the performance of the ancillary equipment within the "unit boundary". Other participants in the test may wish to include or exclude ancillary items simply in order to have the unit conform to their particular definition of a boiler unit. Both viewpoints are understandable, and between countries or standards organizations there may be differences between the compromises adopted. In the final analysis and especially from the boiler purchaser's point of view, as long as the method of calculation is theoretically sound, any differences in declared test results arising merely from different definitions of a unit are of minor importance provided that the basis for obtaining those results is fully understood.

2.2   Closed loops

Another factor which may be relevant when defining the boundary of a unit is the matter of the energy flows in and out of it. In the heat account, the sum of the energy flows inward must equal the sum of the energy flows outward, but some energy flows may be deemed to be "closed loops". These occur when a flow of energy taken from the unit is looped back for use within the boundary of the unit itself. Typical examples are the use of steam for atomizing oil fuel, or for cooling the grate bars of a mechanical stoker.

There are two schools of thought on how such loops should be treated when testing a boiler. One argues that as the operator is concerned only with the net output available for process or heating work, the energy looped back into the unit should be deducted from the total output. The other contends that it is unfair to penalize the apparent performance of a unit by deducting quantities of generated heat output for use on equipment which, on another and otherwise identical unit, would receive energy from an external source. Even in the case of electrically driven ancillaries, the same considerations apply when the unit is used for power generation because it can then be argued that the electrical energy used for driving fans and other equipment comprises part of an external closed loop which returns energy derived originally from the unit.

There is no absolute right or wrong in these different approaches. Provided the basis for assessment of thermal efficiency is agreed and understood by the parties involved in the test, then whether unit boundaries are drawn to include or exclude closed loops is a matter for them alone. This is particularly the case when the test is only of internal interest. Obviously more standardized procedures are inevitable when contractual testing is involved, with the performance of one make or type of boiler possibly being compared with that of another being tested elsewhere. It then becomes especially important for test personnel to follow the recommendations of a particular test code, and it is equally important for the other persons who are comparing test data to bear in mind which code or codes were used, and to understand how variations in those codes could affect the calculations of output and thermal efficiency.

The following notes review the principal ancillaries and energy flows associated with boiler units, and indicate how certain test codes deal with them.

2.3   Ancillaries and energy flows

2.3.1   *Combustion equipment*

The most obvious addition to the heat-exchange sections in a typical unit is the burner, stoker, or other firing equipment. Leaving aside electric boilers, the efficiency of this equipment is something below 100%, although even with solid fuel it might at best be expected that 98% of the calorific value will be released in the furnace. However, it is conventionally accepted in all test codes that the combustion equipment should be included as part of the unit. The matter does not entirely end there, because in some situations the question arises as to how much of the combustion equipment is within the boundary, for example, whether such items as fuel oil heaters or pulverizing mills should be included or not. These matters are considered in 2.3.9. and 2.3.10.

2.3.2   *Feed or return water*

Wherever the boundary is drawn round a unit, the sum of the energy flows crossing it into the unit must equal the sum of the energy flows crossing it in an outward direction. Among the incoming energy flows, the energy in the feed or return water is never, in any test code, treated in the heat account as being part of the heat input.

There are two reasons for this. In many installations the heat in the feed or return flow is simply a proportion of the heat output from the unit itself. It therefore forms part of a closed loop, even though it is largely outside the boundary of the unit. The energy in the feed or return water simply recirculates and cannot be credited to either input or output. Thus, it is normal practice to deduct the heat in the feed water from the heat output.

A different situation can be visualized in which the water fed to the unit might contain heat supplied partially or wholly from a separate source. Even here there would be no point in treating the heat in the feed as part of the heat input. The whole purpose of operating the unit is to transfer heat to the heat carrier and in the case of feed heat, it is already present in the carrier. In other words, such heat is totally unaffected by unit operation or efficiency and could be present whether the unit was operating or not. Again, therefore, the feed or return heat is deducted from the heat output, so that only the heat introduced into the heat carrier within the boundary of the unit is treated as heat output.

2.3.3.   *Mechanical energy of ancillaries*

Any mechanical devices such as pumps and fans which are included within the unit boundary receive supplies of energy to provide motive power, and a proportion of this must eventually be transferred to the fluids flowing within the unit. When this energy is supplied from an outside source it represents an addition to the total energy supplied to the unit. At one time this additional energy was always relatively small, and this is still true in many cases. However, some units do have fans and circulating pumps of relatively high power and it becomes necessary for the sake of accuracy to take this extra input of energy into account.

The question of how much of the total energy supplied should be credited to the heat input is perhaps debatable. It could be argued that in this respect any machine included within the unit should be treated in the same way as the firing equipment, with the entire quantity of energy supplied treated as heat input, and not just the amount subsequently transferred to the fluids. If this were done it might provide further impetus to designers to maximize the efficiency of all ancillaries. However, the convention is for only the shaft powers of ancillaries lying within the unit to be included in the heat account. Thus, the boundary of the unit is assumed to cut across the drive shaft of any mechanical ancillary lying within the unit, and when the power for driving the ancillaries comes from an outside source, only the shaft power of any mechanical ancillary is credited to the heat input. This procedure is followed in both Ref. 4* and Ref. 7. Ref. 4 suggests that if the total amount of mechanical energy supplied by ancillaries amounts to 0.25% or more of the heat input, then it should be taken into account.

Closed loops may be present when the unit itself supplies the energy for driving the machines, either directly in an internal closed loop in the case of steam-driven ancillaries, or via a form of external closed loop when electrical power from a generator supplied with steam from the unit is fed back to drive the ancillaries. In these cases Ref. 4 requires that the heat equivalent of the shaft power be deducted from the heat output but allows other arrangements to be made by contractual agreement. Ref. 5, which applies to units of the power-station type, follows a convention preferred by the British electricity supply industry in which any recuperation of mechanical energy is disregarded. In the direct method the contribution made by ancillary power to thermal input is ignored although some proportion of this power must be present in the measured output and cause the calculated value of thermal efficiency to be higher than the true value. In the indirect method (input unknown) correlation with the direct method is obtained by introducing the ancillary power into the heat account as a "negative item". This, in effect, attributes the ancillary power to output, thus increasing the value of the calculated thermal efficiency to that obtained in the direct method. Acceptability of either the direct or indirect procedures in Ref. 5 hinges upon the thermal input from the mechanical power of ancillaries being relatively small, and this is not always bound to be the case.

In Ref. 7, although a number of ancillaries are included within the unit boundary, none of the energy flows required for driving purposes is treated as a closed loop. The sum of the shaft powers is treated as heat input and the whole of the heat leaving the unit is treated as output, without any deductions. A formula is provided for deriving the shaft power of ancillaries from the power input to the driving motors or turbines.

It should be emphasized that whilst only the power supplied to those ancillaries enclosed within the unit boundary affects the thermal performance, the power supplied to other ancillaries associated with the opera-

*See list of References at end of Appendices.

tion of the unit is nonetheless of interest and should be declared in test results.

2.3.4 *Pumps*

Most steam boilers are equipped with a feed pump, or a circulating pump in the case of hot-water boilers. In all test codes these pumps are considered to lie outside the unit boundary for test purposes, even when they are supplied as part of a packaged unit. The powers are usually declared as matters of interest but are not taken into consideration in the heat account. However, their effect upon the pumped water or other medium in raising its temperature due to the expenditure of mechanical energy does of course affect the calculation of thermal efficiency, because it is that increased temperature, $T21$, crossing the unit boundary, which is measured. In the event that such pumps are steam driven by the unit itself, the steam quantity is not part of a closed loop and is not deducted from the unit output. Attemperator spray water pumps are treated in the same way.

Some units are equipped with recirculation pumps, which promote circulation within the unit itself. These are considered to lie within the unit boundary, and the shaft power supplied to them, if from an external source, is included in the heat account. When these pumps are driven by the unit, Ref. 4 requires the heat supplied to be regarded as a closed loop, and the heat equivalent of the shaft power is deducted from the unit output. Ref. 5 does not specifically refer to circulating pumps in this context. Ref. 7 includes the circulating pump within the unit boundary, but the energy supplied to it is treated as being from an external source and is not deducted from the heat output from the unit.

2.3.5 *Fans*

In its passage to the combustion equipment the air for combustion may pass through a forced draught (FD) fan, or fans, and/or after passing through the heat transfer section of the unit the flue gases may be handled by an induced draught (ID) fan, or fans. In earlier times the energy input to a unit from fans was low, and could reasonably be ignored. In many instances they were not fitted at all, because chimneys were used to provide the moderate draught requirements. This is seldom the case at the present time, when units are usually designed to give high output from compact physical dimensions. Furthermore, some recently introduced types of combustion equipment such as fluidized bed appliances require exceptionally high fan powers. Depending therefore upon where the unit boundary is drawn, fan shaft power may contribute to thermal input. Strictly, only the fan power expended within the unit boundary should be included, and that proportion of the total fan power required to supplement the effect of natural draught in propelling the flue gas through the ductwork beyond the unit and up the chimney should be excluded, but this correction is not made in test codes.

In British test codes the FD fan is considered to lie within the unit boundary because datum temperature is taken to be the air temperature at the fan intake. In Ref. 4 the ID fan may be considered to be either inside or outside the unit boundary (Fig. 2.1). This allows for variations in unit geometry which may affect the taking of the flue-gas temperature (T2) and analysis. It is preferable that these be obtained before the ID fan to eliminate any chance of them being affected by air inleakage at the fan. As the point of measurement of T2 is on the boundary line, then in these circumstances the ID fan is thereby placed outside the boundary. However, some units are designed with built-in ID fans, and in such cases it is impossible to obtain instrumentation points before the fan. There is then no alternative but to have the unit boundary extending beyond, and including, the ID fan.

American practice (Ref. 7) places both FD and ID fans outside the unit boundary, and their shaft powers are not included in the heat account. This test code also allows for the presence of air heaters, external to the unit, and situated between the FD fan and the unit boundary.

2.3.6   *Datum temperature*

A fundamental factor when determining the magnitude of sensible heat flows crossing the boundary of the unit is the datum temperature upon which they are based, and this temperature is identified with a specific location on the unit boundary in a number of test codes. Thus Refs. 3, 4 and 5 take as datum the temperature of the air at entry to the unit (normally entry to the FD fan). This is assumed to be the ambient temperature in the vicinity of the unit and is also the temperature on which all other heat flows into and out of the unit are based.

In Ref. 7 the test is based on a reference air temperature which is a standard or guarantee condition, and not necessarily that obtaining at the time of the test. In fact, the document allows in any case for the existence of air heaters between the FD fan (considered to lie outside the unit boundary) and the unit itself, so allowance is made in the test calculations for the heat in the air for combustion, and also in the humidity moisture in that air, at entry to the unit. A similar approach is adopted in Ref. 5 to allow for preheating of the air for combustion, or variations in ambient air temperature. See also the references to datum temperature in 2.3.11.

2.3.7   *Steam injected into furnace*

Steam may be injected into the furnace of the unit for various reasons — to atomize oil, to create turbulence for promoting combustion, and to cool the grate bars of mechanical stokers (alternatively, water may be injected for this purpose).

Ref. 4 allows for the associated heat flow to be either from an outside source or from the unit itself, and in the latter case the heat in the steam is deducted from the heat output whilst the subsequent loss in the flue gas due to the extra moisture is allowed for in the heat account. In Ref. 5 the heat in the steam is deducted from the heat output, and no distinction is made between steam supplied by the unit itself and steam supplied from an outside source.

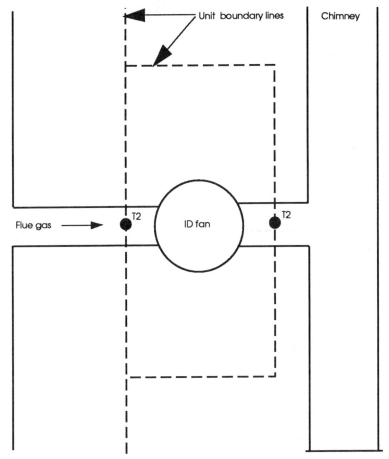

Fig. 2.1   ID fan — alternative unit boundary lines

In Ref. 7 the boundary is drawn so that the steam is treated as being always from an outside source, and no heat is deducted from the heat output.

2.3.8   Blowdown

The flow of blowdown water from a unit may be sufficiently large to contain a significant amount of heat. Even when the heat or any portion of it is not recuperated, both British and U.S. practice in tests is to refrain from blowing down during the test or to credit the heat in the blowdown to the unit output. There are two reasons for this, both stemming from the fact that test codes are particularly concerned with contractual tests.

(a) It can be argued that as the amount of blowdown depends upon the quality of the feed water supplied to the unit, then if the heat in blow-down were treated as a loss, two otherwise identical units would, at different sites and with different qualities of feed water, yield different thermal performance figures.

(b) Measurement of blowdown quantity may be difficult, or discharge may be intermittent and infrequent, so test codes allow operation of blowdown equipment to be suspended over the period of a test. Again, this would mean that if blowdown were counted as a loss, a unit equipped with an automatic continuous blowdown system would be at a disadvantage, compared with an otherwise identical unit in which the operation could be suspended.

Needless to say, such considerations as these do not apply when assessing the thermal performance of a unit in its everyday work. Any loss occurring in blowdown must be recognized as such, and if it is considered unacceptably high, corrective action can be taken by improving the quality of the boiler feed water.

2.3.9   *Heating of fuel oil*

The heavier grades of fuel oil are heated for storage and handling to temperatures between 10 and 55°C, and then to give the required low viscosity (18–25 cSt) for atomization they are further heated up to perhaps 250°C. The high-temperature or secondary heater is in close proximity to the unit and in British practice is considered to lie within the unit boundary.

Ref. 3 includes all of the sensible heat in the oil in the heat input. Ref. 4 distinguishes between the condition when all of the heat in the oil is supplied from an outside source (and is taken in total to be an addition to the heat input to the unit), and the condition when the unit itself supplies the heat (Fig. 2.2). In this case the heat supplied to the secondary heater is treated as a closed loop and deducted from the heat output, with the remaining heat in the oil treated as a heat input. Even if the unit is supplying the energy for heating the storage tank and pipework, this is not deducted from the unit output as it would clearly be unfair to reduce the apparent performance of the unit because of site layout conditions. It is after all possible to imagine a situation where the oil installation was so widely spaced that the entire output of the unit was used in heating it.

In Ref. 5 the heat in the oil between the temperature at which it is weighed and the temperature at the burner is deducted from the unit heat output without distinction being made between the case where the heat is supplied from an outside source and where it is supplied by the unit itself.

In Ref. 7, the secondary oil heater is considered to lie outside the unit, so all of the heat is treated as an extra heat input and no deduction is made from the unit heat output.

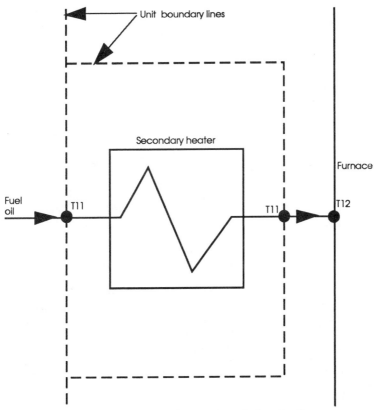

Fig. 2.2   Fuel oil secondary heater — alternative boundary lines

2.3.10   *Coal pulverizing mills*

When a solid fuel is fired by pulverized fuel (PF) in which coal is ground so that — typically — 80% will pass through a 76 micron mesh, the PF may be

(a) milled on site and used directly;
(b) milled on site, put into storage, and drawn off as required for combustion;
(c) delivered to the site ready milled.

Obviously there is much more energy used on site when the milling operation is carried out there and so, to promote fairness in comparison between similar units operating on PF from different sources or with mills having different efficiencies, the convention in Britain (Refs. 4 and

5) is for the pulverizing mills to be considered as being outside the unit boundary, although in the case of Ref. 5 this point is not made specifically. No specific recommendations are made concerning the treatment of the energy supplied to the mills in the case of method (a), although part of this energy must contribute towards the heating of the air and coal passing to the unit. This quantity of energy may be appreciable, and especially so when "microfine" PF is being produced. This form of PF, in which 90% of the milled coal will pass through a 40 micron mesh or less, has attracted interest as a means of converting oil-fired plant to solid fuel. The energy consumption required for milling is significantly higher than is the case with ordinary PF and when fluid-power mills are used, in which high-velocity jets of steam cause the coal particle sizes to be reduced by attrition, up to 8% of the entire steam output of a unit may be used in the mills. Following British practice, the heat in the steam would not be deducted from the unit output but would presumably be added to the heat input. Similarly, it might be expected that the mechanical energy supplied to any PF mill when it is electrically driven should, in British practice, be added to the heat input, but this is not done.

In America (Ref. 7) the mills are included within the unit boundary and the energy supplied to drive the mills is treated as an addition to the heat input.

In both British and American test codes (Refs. 4, 5 and 7) the heated air supplied by the unit to the PF mills for drying the coal is treated as a closed loop, and the heat content is therefore ignored in the heat account.

It is to be noted that in the case of on-site milling the chemical analysis of the coal will normally be that of the raw coal prior to milling, and therefore in the test calculations British practice (Refs. 4 and 5) requires some correction to be made if any significant quantity of material is rejected by the mill. In American practice (Ref. 7), as the mill is inside the unit boundary, the rejected material is treated in the same way as other solid residues.

2.3.11   *Radiation*

A loss which inevitably occurs when a unit is operating is that due to radiation, conduction and convection from its heated surfaces. These surfaces are at or near normal working temperatures regardless of the level of output, so it follows that the rate of heat loss is approximately constant and therefore, when declared as a proportion of heat input, the radiation loss is lowest when the unit is working at maximum output. When a unit is situated in a boilerhouse, it follows that some of the heat radiated from the boiler is recuperated in heating the combustion air before it enters the unit. This recuperation constitutes another closed loop, as Fig. 2.3 illustrates.

If it is imagined that all of the heat radiated from the unit (L85) is passing through a heat exchanger, some (L86) eventually passes out to atmosphere through the boilerhouse structure. The remainder (L85 − L86) is returned to the unit in the combustion air, which is raised

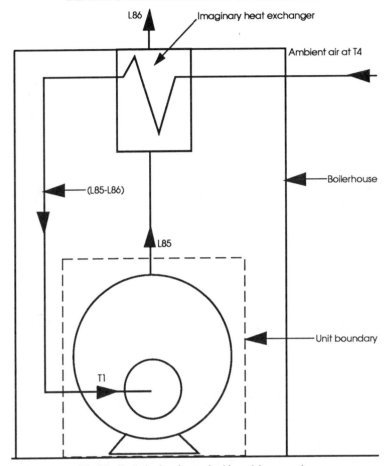

Fig. 2.3   Radiation loss from unit with partial recuperation

in temperature from the outside ambient temperature T4 to the temperature at the inlet to the unit, T1.

Note, however, that when T1 is taken as datum temperature, recuperation is discounted and disappears from the heat account. The radiated heat from the unit which appears in the heat account is then the total heat radiated to the boilerhouse, not the net heat after allowing for recuperation as is sometimes thought.

It might be argued that as all air supplied to the unit is originally at temperature T4 outside the boilerhouse, this would be a better datum, but in the case of an indoor unit the flue gases could not in practice be

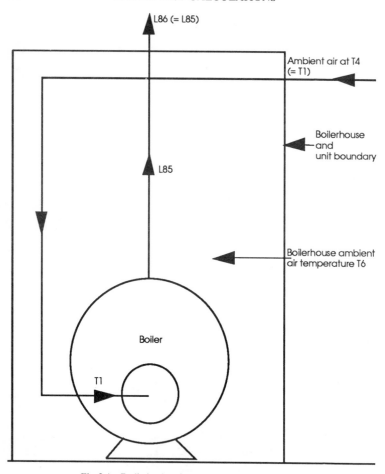

Fig. 2.4   Radiation loss from unit without recuperation

cooled to a lower temperature than the boilerhouse ambient temperature, so T1 is a more realistic datum. Furthermore, as Fig. 2.3 shows, because the boundary would then be represented by the boilerhouse structure, the radiated heat which would appear in the heat account would be that leaving the boilerhouse, not the boiler, that is, L86. If air from outside the boilerhouse is ducted directly to the unit as in Fig. 2.4, T1 is equal to T4, no recuperation takes place, and the total radiation loss from the boiler (L85) is equal to the radiation loss from the boilerhouse (L86).

It should be noted that whilst the temperature at entry to the unit is often taken as datum, correction may be made to the test calculations in

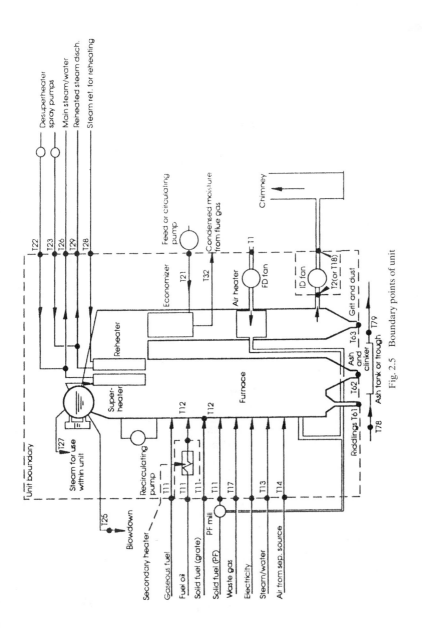

Fig. 2.5   Boundary points of unit

order to have these based upon some reference temperature which has been selected by agreement. This is featured in the procedure in Ref. 7.

2.4   Boundary points of unit

The previous notes will have served to show that, particularly in the case of more complicated units, it is essential to decide where the boundary of the unit is assumed to be prior to any test taking place.

Whilst it would be impossible to attempt to give guidance on every possible present or future boiler design variation, Fig. 2.5 indicates the principal heat flows that may be encountered on a conventional unit. Naturally, it would be highly unlikely that all of them would be found together on any particular unit. Refs. 4 and 7 provide diagrams of this type for guidance.

3. Fuels

3.1 Fuel types

In carrying out the calculations associated with detailed thermal tests it is necessary to know the chemical analysis of the fuel, although in simplified test procedures a partial analysis and/or constants relevant to general types of fuel may be used.

The fuels used in boiler furnaces may be described broadly in terms of their state, that is, as solid, liquid or gaseous. Electricity, when used as a boiler "fuel", poses none of the problems associated with the primary fuels as far as derivation of heat input or flue-gas losses are concerned, and does not therefore warrant detailed consideration.

Common to all of the fossil fuels are the basic fuel elements, hydrogen and carbon. In liquid and gaseous fuels these are wholly combined into hydrocarbons of one sort or another. In coal, some hydrocarbons are present, but the majority of the carbon is uncombined or "fixed". Sulphur is present in many fuels and has some calorific value, but is always regarded as a contaminant.

Detailed specification of fuels involves numerous factors which have an important influence upon boiler operation, but only a few which have a bearing upon the test calculations. The following notes are intended for general guidance but have been extended a little beyond the most basic information necessary for test calculations, so that references in other sections to fuel types and particular characteristics may be understood by those who are unfamiliar with any of the principal fuels.

3.2 Solid fuel

Solid fuels include those which occur naturally such as coal, lignite, brown coal and peat, and also a number of wastes such as town refuse, wood waste, bagasse (the fibrous residue from sugar cane), nutshells, straw and so on. Coke, and also the various briquetted fuels which are manufactured from coal, are widely used in domestic heating appliances because they burn smokelessly, and special cokes are used in metallurgical and other furnaces, but such fuels are not normally used for firing industrial or power-station units because of their relatively high cost. In mainland Britain, where there is little lignite or brown coal, the most familiar of the solid fuels used in industry and for power generation are anthracite, low-volatile, and bituminous coals (substantial deposits of lignite exist in Northern Ireland but have not been exploited at the time of writing). In America, a wider range of coals is available.

The following terms are particularly relevant in describing coals, though many of them are also applicable to solid fuels in general.

3.2.1   *Volatile matter*

This is the loss in mass, less that due to moisture, which appears when coal is heated to 900°C out of contact with air under standardized conditions. Principally, it comprises a mixture of tarry vapours and combustible gases such as hydrogen and the hydrocarbon gases methane, ethane, and so on.

Coals with high proportions of volatile matter burn more readily than those with less and are described as "free burning" or "reactive" fuels, but they require more careful adjustment of fuel and air flows to avoid smoke emission.

3.2.2   *Moisture*

Coal as delivered will normally contain water as a proportion of its mass. If a sample of this coal is allowed to dry by exposure to the atmosphere in a well-ventilated place any resulting reduction in mass arises because of the evaporation of free or surface moisture. This air-dried coal still contains a further quantity of inherent moisture absorbed within the structure of the coal itself. The total moisture is the sum of both free and inherent moisture contents. In boiler test calculations the total moisture must be known in order to calculate the loss due to water vapour in the flue gas.

In considering the advantages or disadvantages of any coal as a fuel, the free moisture, considered together with the proportion of $< \frac{1}{2}$ mm coal particles, is important. Difficulties in handling may be acute if there is about 12% free moisture associated with 10–20% fine particles below $\frac{1}{2}$ mm. On the other hand, when burning coal with a high content of < 3 mm particles on mechanical stokers some free moisture is desirable as it assists in binding the particles together and thereby prevents a high thermal loss due to unburnt particles either falling through the stoker grate bars or being carried out of the units in the flue gas.

3.2.3   *Ash and mineral matter*

Ash is the solid residue remaining after the combustion of coal, but the term is also used generally in relation to the inert material in the unburnt coal which will subsequently comprise the ash. This inert material, largely silica, alumina and ferric oxide, but with many other elements in smaller quantities, may be present as inherent ash, that is, as an integral part of the coal structure. However, it may also be present as extraneous ash, having been mined from mineral strata adjacent to the coal seam. Such ash can be largely removed at a coal preparation plant by washing or other cleaning processes.

"Mineral matter" is a calculated quantity, slightly larger than the ash quantity, which is required when making precise comparisons between coals. It allows for the actual behaviour of inert material in the combustion process which is used in determining the proportion of "ash", for example, the volatilization of certain matter and the retention in the residues of some sulphur. In Britain the mineral matter is derived using the King Maries Crossley formula, which requires a detailed chemical analysis of the coal; or alternatively, if the amount of analytical data is more limited, a simplified formula devised by British Coal, then the National Coal Board, in 1956 may be used. In America, a formula devised by Parr is used.

In boiler test calculations, ash is always a significant factor, most particularly because it will inevitably contain a proportion of unburnt combustible matter, but also because if it is rejected at a significantly high temperature, the resulting sensible heat loss will also have to be taken into account.

The physical and chemical characteristics of the ash are important in deciding upon the suitability of a coal for a particular application. In some combustion appliances it is necessary to remove the ash as discrete particles, so an ash with a high fusion temperature is preferred. In others it is more convenient to allow the ash to fuse and remove it as a clinker or even as molten slag, so ash having a lower fusion temperature is required. Certain elements may also tend to promote the formation of deposits on furnace or convection surfaces, thereby altering the performance of the boiler.

3.2.4   *Analysis*

Coal is naturally variable in composition, whilst mining, preparation, transport and stocking methods can affect the ash and moisture contents and hence the overall analysis and calorific value. A number of analyses may be referred to in relation to coal. Thus, an "ultimate" analysis is an analysis specifying the proportion by mass of carbon, hydrogen, oxygen, nitrogen, sulphur, ash, and moisture. It represents the essential information necessary for a detailed test on a unit. A "proximate" analysis is a mass analysis in terms of fixed carbon (i.e. carbon which does not pass off in volatile matter), volatile matter, ash and moisture. It provides a practical basis for comparison of fuels in terms of broad characteristics and likely performance on combustion equipment. In more simplified boiler tests, this is sometimes the only analysis declared, although it is then necessary to augment it, either directly with the hydrogen content or alternatively with some means of deriving this.

The fuel analysis required for test purposes is that applying to the fuel as it actually enters the unit, and in the case of coal or any other solid fuel this may or may not be a convenient point to take samples. It is, in particular, the free or surface moisture content of coal that can vary, even within a given plant, as the coal is moved from point to point, so allowance must be made for any moisture lost or added between the sampling point and entry to the unit. A number of analyses relating to coal result from the variations which can occur simply from the specified location, for example, "as received", "as weighed", and "as fired". Moisture losses may occur through evaporation or drainage, whilst additions may occur through exposing stock to rain, or by deliberately adding water in measured amounts to smalls coals (see 3.2.7) which are exceptionally dry. As explained in 3.2.2, this is done to promote combustion efficiency. There are, too, specially prepared coal–water mixtures in which finely ground coal comprises about 70% of the total mass, the remainder being almost entirely water, with small amounts of additives to promote the formation and maintenance of a homogeneous fluid. These mixtures are considered to have potential in the conversion of oil-fired plant to coal, or in other applications such as marine use, where ease of fuel handling is of particular importance.

Another condition pertaining to the level of moisture is "air dried". This refers to coal which has been exposed in a dry, well-ventilated environment and allowed to lose its free or surface moisture. It is the condition to which coal samples are brought prior to laboratory analysis. The loss of weight due to the evaporation of free moisture is measured to enable suitable correction to be made to the laboratory "air-dried" analysis to obtain the "as-sampled" analysis. "Dry" is a term applied to a totally moisture-free condition, that is, with both free and inherent moisture removed. In America the term "moist" is applied to coals containing only inherent moisture.

Ash is another constituent which may vary from one analysis to another, and so in order to get a firm basis for the comparison of coals by restricting consideration to the coal substance only, coal analyses are also declared on a "dry, ash-free" (d.a.f.) basis. The "ash" in this case is the quantity determined in direct analysis. In pursuit of even greater precision, the "dry, mineral-matter-free" (d.m.m.f.) basis is used and is the one upon which, for example, British coals and the higher-rank American coals are classified. As stated under 3.2.3, the analysis of the d.m.m.f. basis makes allowance for the changes which occur in the mineral matter during combustion and which result in the calculated dry mineral matter being slightly greater in mass than ash as directly determined.

The method of converting from one analytical basis to another is described in Appendix 4.

3.2.5   Caking, coking and swelling

When coal particles are heated, any tarry matter which evolves tends to bind adjacent particles together, and this enables the mass of coal to achieve a plastic state. Following further heating and devolatilization, a cake of agglomerated particles is formed, and it is this process which is the essential factor in the production of a coke. Associated with this heating and caking there may also be the tendency of the coal to swell. In Britain, the ability of coals to cake is measured in the Gray–King coke tests, in which samples of powdered coal are heated and the profile of the carbonized residue is compared with standard profiles, classified by a range of letters from A, which is non-caking, to G, which is strongly caking. The more highly swelling coals are classified by an extension to the range in which numbers from 1 upwards are added to the G classification (Ref. 12). Swelling is measured in terms of crucible swelling numbers. This is a scale of numbers from 1 to 9, rising in half units. They are allocated by comparing coke buttons produced under test with standard profiles (Ref. 12). The same system is used in America (Ref. 17).

3.2.6 Rank and classification

The rank of coal is a measure of its maturity, or carbon content, and it is upon this basis that coal is classified. Anthracite is the most mature of all the coals, has the highest carbon content, and is therefore of the highest rank. Below it come semi-bituminous coals, bituminous coals, lignite, brown coal, and peat. Peat would not normally be described as a coal, but represents an early stage in its formation.

Several classification systems are in use. The one used in Britain for indigenous coal, Table 3.1, is that devised by British Coal and classifies the coal within a range of Coal Rank Code numbers extending from 100 to 900, with the lowest numbers indicating the highest rank. These numbers are allocated in accordance with the volatile matter on the d.m.m.f. basis, and the coking characteristic as defined by the Gray–King Assay. In general, the volatile matter increases as the rank decreases (that is, as the Coal Rank Code number increases towards 900) whereas the caking property is zero at both ends of the scale and is at a maximum at intermediate ranks.

Four main classes are identified in the system, namely, anthracite (class 100), low-volatile steam coals (class 200), medium-volatile coals (class 300) and high-volatile coals (classes 400–900). Classes 100–200 are subdivided on the basis of volatile content alone, as they have negligible caking properties. In some cases coals have had their original characteristics modified by subsequent geological events and suffixes are used to identify these. Thus, heat-altered coals which have been affected by igneous intrusions are denoted by "H" and coals which have been oxidized by weathering carry the suffix "W".

In America the classification system includes four principal classes, anthracite, bituminous, sub-bituminous, and lignitic coals (Ref. 17). Each class is divided into a number of groups. In the case of higher-rank coals the groups are defined in terms of the fixed carbon and volatile matter contents on the d.m.m.f. basis. Groups in the lower-rank coals are defined in terms of the calorific values on the moist mineral-matter-free basis, where "moist" refers to inherent moisture only. Elements from both these methods of definition may be applied to coals in the middle of the bituminous class. The rank classification is expressed in the form of two numbers, the first being the percentage of fixed carbon on the dry basis to the nearest whole number. The second is the calorific value on the moist basis expressed as number of hundreds of Btu/lb to the nearest hundred. The classification is shown in parentheses to indicate that it is on the mineral-matter-free basis. The classification may be followed by other number's specifying, in abbreviated form, the factors which constitute the grade of an American coal. See 3.2.7.

From the point of view of combustion, a high-rank coal such as anthracite has a high calorific value. It is naturally smokeless because of its low volatile content, although it is less reactive for the same reason. High-volatile coals burn easily but require careful balance of fuel and air flows if smoke emission is to be avoided. The highly caking coals may be unacceptable for use on certain appliances because of their tendency to cake and impede the flow of combustion air. On other appliances and with appropriate techniques, they are quite satisfactory.

3.2.7   *Preparation and grading*

When it is first brought to the surface from a deep mine, "run of mine" coal, as produced by modern mining machinery, comprises a wide range of particle sizes and also includes quantities of extraneous mineral matter removed from strata above and below the coal seam. Where site conditions permit, use of open-cast or strip-mining techniques may result in the production of cleaner coals which include less extraneous mineral matter. At a preparation plant near the mine, various washing processes may be used to reduce the amount of mineral matter and therefore a coal may be available as untreated, washed, or possibly marketed as a blend of the two ("part treated"). The same preparation plant makes use of screens to produce a number of specific sizes to suit market requirements. In Britain these are referred to as grades when they have both upper and lower size-limits, and as smalls, duff or fines when they have only an upper size-limit. There are also "large" coals which have only a lower size-limit specified. The grades marketed in Britain by British Coal are listed in Tables 3.2 and 3.3. In America, "grade" is a more general description which specifies size, calorific value, ash content, ash softening temperature, and sulphur content. Coals which have their sizes designated in terms of upper and lower limits are referred to as "double screened" (Ref. 17).

As examples of the uses of coals, in Britain large amounts of coal are burnt as PF for power generation, and a typical power-station coal is a blend of washed and untreated smalls having a top size of 50 mm or less and an average ash content of 17%. Large industrial boiler plants are usually supplied with washed or blended smalls, having a top size of about 25 mm, and an ash content of perhaps 7–10% or more. Smalls are the preferred fuel for certain appliances such as the chain-grate stoker but also have the commercial advantage of being lower in cost than graded coals. Smalls are not so easily handled as graded coals and for this reason would not usually be specified for small plants. In smaller industrial plants, washed singles are widely used, having an ash content of about 5%. This is the preferred fuel for such appliances as the underfeed stoker but also has the general advantage when used on any type of stoker of being easily handled.

Table 3.1   The British Coal coal classification system (revision of 1964)

Coal Rank Code			Volatile matter (d.m.m.f.) (%)	Gray–King coke type*	General description
Main class(es)	Class	Sub-class			
100			Under 9.1	A	Anthracites
	101†		Under 6.1	A	
	102†		6.1–9.0		
200			9.1–19.5	A–G8	Low-volatile steam coals
	201		9.1–13.5	A–C	
		201a	9.1–11.5	A–B	Dry steam coals
		201b	11.6–13.5	B–C	
	202		13.6–15.0	B–G	
	203		15.1–17.0	E–G4	Coking steam coals
	204		17.1–19.5	G1–G8	
300			19.6–32.0	A–G9 and over	Medium-volatile coals
	301		19.6–32.0	G4 and over	
		301a	19.6–27.5 }	G4 and	Prime coking coals
		301b	27.6–32.0 }	over	
	302		19.6–32.0	G–G3	Medium-volatile, medium-caking or weakly caking coals
	303		19.6–32.0	A–F	Medium-volatile, weakly caking to non-caking coals
400–900			Over 32.0	A–G9 and over	High-volatile coals
400			Over 32.0	G9 and over	High-volatile, very strongly caking coals
	401		32.1–36.0 }	G9 and over	
	402		Over 36.0 }		
500			Over 32.0	G5–G8	High-volatile, strongly caking coals
	501		32.1–36.0 }	G5–G8	
	502		Over 36.0 }		
600			Over 32.0	G1–G4	High-volatile, medium caking coals
	601		32.1–36.0 }	G1–G4	
	602		Over 36.0 }		
700			Over 32.0	E–G	High-volatile, weakly caking coals
	701		32.1–36.0 }	E–G	
	702		Over 36.0 }		
800			Over 32.0	C–D	High-volatile, very weakly caking coals
	801		32.1–36.0 }	C–D	
	802		Over 36.0 }		
900			Over 32.0	A–B	High-volatile, non-caking coals
	901		32.1–36.0 }	A–B	
	902		Over 36.0 }		

NOTES

Coals with ash of over 10% must be cleaned before analysis for classification to give a maximum yield of coal with ash of 10% or less.

Coals that have been affected by igneous intrusions ("heat-altered" coals) occur mainly in classes 100, 200 and 300, and when recognized should be distinguished by adding the suffix H to the Coal Rank Code e.g. 102H, 201bH.

Coals that have been oxidized by weathering may occur in any class, and when recognized should be distinguished by adding the suffix W to the Coal Rank Code, e.g. 801W.

*Coals with volatile matter of under 19.6% are classified by using the parameter of volatile matter alone: the Gray–King coke types quoted for these coals indicate the general ranges found in practice, and are not criteria for classification.

Notes continued next page

†In order to divide anthracites into two classes, it is sometimes convenient to use a hydrogen content of 3.35% (d.m.m.f.) instead of a volatile matter of 6.0% as the limiting criterion. In the original Coal Survey rank coding system the anthracites were divided into four classes then designated 101, 102, 103 and 104. Although the present division into two classes satisfies most requirements it may sometimes be necessary to recognize more than two classes.

Technical Data on Solid Fuel Plant (Ref. 14)

Table 3.2   Sizes of British Coal graded bituminous coals

Description	Method of preparation
Large cobbles	Prepared through screens not less than 150 mm (6 in.) and over screens not less than 75 mm (3 in.).
Cobbles	Prepared through screens not less than 100 mm (4 in.) but not greater than 150 mm (6 in.); and over screens not greater than 100 mm (4 in.) but not less than 50 mm (2 in.).
Trebles/Large nuts	Prepared through screens not less than 63 mm (2½ in.) but not greater than 100 mm (4 in.); and over screens not greater than 63 mm (2½ in.) but not less than 38 mm (1½ in.).
Doubles/Nuts	Prepared through screens not less than 38 mm (1½ in.) but not greater than 63 mm (2½ in.); and over screens not greater than 38 mm (1½ in.) but not less than 25 mm (1 in.).
Singles	Prepared through screens not less than 25 mm (1 in.) but not greater than 38 mm (1½ in.); and over screens not greater than 18 mm (¾ in.) but not less than 12.5 mm (½ in.).

NOTES
All references to screens are to commercial screens with round holes of diameters as shown or other equivalent apertures.
Where good screening practice requires, British Coal may include in first sections of the screens some plates having apertures outside the above ranges. This is necessary on grounds of underflow control, screen balance and effective screening having regard to particular coal characteristics.
British Coal reserve the right to change at any time and without notice the sizes of screen apertures in use within the ranges shown.

Technical Data on Solid Fuel Plant (Ref. 14)

Table 3.3   Sizes of British Coal graded low-volatile coals

Welsh anthracite			Welsh dry steam coal		
Name	mm	in.	Name	mm	in.
			Cobbles	125/100–83	$5/4$–$3\frac{1}{4}$
Large nuts	75–50	3–2	Large nuts	83–50	$3\frac{1}{4}$–2
Small nuts	50–22	2–$\frac{7}{8}$	Small nuts	50–19/17.5	2–$\frac{3}{4}/\frac{11}{16}$
Beans	22–11	$\frac{7}{8}$–$\frac{7}{16}$	Peas	19/17.5–9.5	$\frac{3}{4}/\frac{11}{16}$–$\frac{3}{8}$
Grains	11–6.3	$\frac{7}{16}$–$\frac{1}{4}$			
Washed duff	4.75–0	$\frac{3}{16}$–0	Washed duff	12/9.5–0	$\frac{1}{2}/\frac{3}{8}$–0

Technical Data on Solid Fuel Plant (Ref. 14)

3.3 Liquid fuel

Any liquid having calorific value may be used as a fuel, but in general the liquid fuels comprise almost entirely those petroleum products which are produced specifically for that use.

In the past, a range of liquid fuels derived from coal tar was in use in British industry. These "Coal Tar Fuels", or CTF, derived from coal carbonization and gasification processes and specified in Ref. 8, were available in six grades, two being distillates with the remainder of heavier type.

Some years ago when doubts concerning the availability and escalating price of fuel oils led energy consumers to examine alternatives, coal/oil mixtures were investigated. Finely pulverized coal comprised some 20–30% of these mixtures, but the cost of preparation and the fact that oil still comprised the bulk of the fuel discouraged further progress.

The following notes outline the principal characteristics underlying the classification of petroleum fuel oils.

3.3.1 Hydrocarbon constituents

The fuel elements in fuel oils are hydrogen and carbon, combined together into numerous hydrocarbon compounds. These fall into several categories, according to their molecular structure. Thus, the "aliphatic" compounds — paraffins and olefins — have the carbon and hydrogen atoms arranged in chains or branched chains, whilst the aromatics and naphthenes have at least some of their carbon atoms arranged in rings.

3.3.1.1 Paraffins

These include methane, propane, and so on, and have the general formula $CXHY$, where $Y = 2X + 2$. They usually comprise most of the mass of the distillate fuels and have the following characteristics:

(a) They have the lowest carbon content of any hydrocarbon and so require most air for combustion, but they are least likely to produce smoke or soot if combustion conditions are incorrect.
(b) Some paraffins have high freezing-points, so it is these hydrocarbons which are likely to cause thickening of oil fuels at low temperatures.

3.3.1.2 Aromatics

These include benzene and toluene and are so called because some of their derivatives have aromatic odours. They have the general formula $CXHY$, where $Y = 2X - $ (a number dependent upon molecular size), and the following characteristics:

(a) Their high carbon content requires careful adjustment of fuel and air flows to avoid emission of smoke or deposition of carbon.
(b) They improve the fluidity of the oil at low temperatures.

3.3.1.3 Naphthenes (or cyclo-paraffins)

These include, for example, cyclohexane and have the general formula $CXHY$, where $Y = 2X$. These compounds have a carbon/hydrogen ratio between those of the paraffins and aromatics.

3.3.1.4   Olefins

These compounds include ethylene and propylene and have the same formula as the naphthenes but occur only as a result of high-temperature cracking processes breaking down large molecules into smaller ones.

3.3.2   *Sources*

As found naturally, crude oil may contain predominant quantities of particular hydrocarbons and as a result may be described for example as paraffinic, aromatic, naphthenic, or asphaltic. Asphaltic crude is the heaviest, and comprises bitumen and inert matter (bitumen being a black or dark brown solid or semi-solid organic material which liquefies when heated).

3.3.3   *Distillate and residual oils*

"Distillate" and "residual" are terms often quoted in the classification of oil fuels and stem from the route taken in production. As the term distillate suggests, an oil of this type has been condensed as one of the lighter fractions of the mixture of vapours produced as a result of heating and evaporating a crude oil. Distillation, sometimes repeated, and other conversion processes such as "cracking" — the breaking of large molecules into lighter fractions — and also the blending of light and heavy oils, are used to produce a range of oils from crude to suit particular market requirements. The heavier residual oils differ from distillates in that they contain a greater proportion of residual matter (and therefore more ash-forming constituents) and require heating for effective handling and atomization.

Recognition of specific hydrocarbon compounds is possible in the lighter distillate fuels, but heavier distillates and residual fuels are highly complex mixtures of vast numbers of different types. This is unimportant in boiler tests, where only the ultimate analysis specifying the whole carbon and hydrogen content by mass is required. Even from the purely practical standpoint of achieving efficient combustion, a broad knowledge of the classification of the oil is all that is required to specify suitable combustion equipment of known capability.

3.3.4   *Viscosity*

Viscosity is a measure of the fluidity, or the internal resistance to flow of a fluid, and may be defined as the force necessary to shear a cube of the fluid at unit velocity. Both burning and pumping characteristics depend on viscosity, and it is the principal factor by which a liquid fuel is classified.

In Britain the characteristics of fuel oils are defined in Ref. 9 for classes designated as C, D, E, F, G and H in ascending levels of viscosity, with class H the heaviest or most viscous at a maximum of 56 cSt at 100°C. In America, fuel oils are graded by numbers that rise with increasing viscosity in the series 1, 2, 4, 5 and 6. The viscosities are declared at temperatures of 100 or 120°F (Ref. 18).

The centistoke is an absolute unit of kinematic viscosity, obtained by measuring the flow rate in a standard U-tube, and may be used in pumping calculations. Several other viscosity units are or have been in use. The Redwood scale of viscosity units was formerly very popular but now its use is firmly discouraged because it is not absolute. Redwood viscosities were established in an arbitrary way by measuring the time taken in seconds for a quantity of oil at a specified temperature to flow through the orifice of a Redwood viscometer. Though not in favour in the scientific sense, Redwood viscosities are still widely, if inaccurately, quoted when referring to oil fuels; for example, a Class G oil may be referred to as a "3500 seconds" oil. Table 3.4 lists the properties of British fuels, whilst Fig. 3.1 shows the relationship between viscosity and temperature for residual fuel oils.

3.4   Gaseous fuels

In Britain the most widely used gaseous fuel is North Sea natural gas, largely methane, which is distributed through an extensive pipework system. This entirely replaced the "town gas" produced originally from coal and then later from hydrocarbon oils.

In large plant, assuming an economic case can be made, gaseous fuels of various types can be generated for local use. Some processes also produce combustible gases as by-products, for example, blast furnace gas in steel works or methane in some collieries, and these too are valuable as fuels for local use.

The fuel gases butane and propane have become generally available as liquefied petroleum gas (LPG), being widely distributed in containers under slight pressure and resuming the gaseous state when released to atmosphere.

3.4.1   Composition

Gaseous fuels comprise either one single fuel gas, or — most usually — a mixture of various gases which may include oxygen, and also inert gases such as nitrogen and carbon dioxide. The analysis is declared on a volumetric basis, listing the proportions of the constituent gases, but if it is desired to use the same test formulae for all fuels then the volumetric analysis is converted to a mass basis and declared simply in terms of the basic chemical elements (see Appendix 6).

3.4.2   Classification

The characteristics of gaseous fuels are conveyed in broad terms by their names, for example, North Sea natural gas, blast furnace gas, producer gas, commercial butane and so on. A more scientific classification is based upon the Wobbe number, which is the ratio of the calorific value of a gas to the square root of its specific gravity. It is proportional to the rate at which heat is supplied when burners are operated at constant pressure.

In Britain, fuel gases are divided into first, second and third "families", with each family defined in terms of a range of Wobbe numbers (Ref. 16). Thus when the Wobbe numbers are based on gross calorific values the first family covers numbers in the range 22.7–29.8 MJ/m^3, the second

Table 3.4 Properties of burner fuels

Property	Class C1	Class C2	Class D	Class E	Class F	Class G	Class H	British Standard	Technically equivalent Institute of Petroleum standard
Viscosity, kinematic at 40°C, cSt								BS 2000: Part 71	IP 71
min.	—	1.00	1.50	—	—	—	—		
max.	—	2.00	5.50	—	—	—	—		
Viscosity, kinematic at 100°C, cSt, max.	—	—	—	8.20 (see C.1)	20.00 (see C.1)	40.00 (see C.1)	56.00 (see C.1)	BS 2000: Part 71	IP 71
Carbon residue, Ramsbottom on 10% residue, % (m/m), max.	—	—	0.2					BS 4451	IP 14/65
Distillation								BS 2000: Part 123	IP 123
recovery at 200°C, % (V/V), min.	15.0	15.0	—	—	—	—	—		
recovery at 200°C, % (V/V), max.	60.0	—	—	—	—	—	—		
recovery at 350°C, % (V/V), min.	—	—	85.0	—	—	—	—		
final boiling point, °C, max.	280	300	—	—	—	—	—		
Flash-point, closed, Abel, °C, min.	43.0	38.0	56.0	—	—	—	—	BS 2000: Part 170	IP 170
Flash-point, closed, Pensky–Martens, °C, min.	—	—	—	66.0	66.0	66.0	66.0	BS 2000: Part 34	IP 34

Property									
Water content, % (V/V), max.	See 3.3	See 3.3	0.05	0.5	0.75	1.0	1.0	BS 4385	IP 74/82
Sediment, % (m/m), max.	See 3.3	See 3.3	0.01	0.15	0.25	0.25	0.25	BS 4382	*
Ash, % (m/m), max.	—	—	0.01	0.15	0.15	0.20	0.20	BS 4450	—
Sulphur content, % (m/m), max.	0.04	0.20	—	—	—	—	—	BS 2000: Part 107	IP 107
	—	—	0.50†	—	—	—	—	BS 5379 (EN 41)	—
	—	—	—	3.50	3.50	3.50	3.50	BS 2000: Part 61	IP 61
Copper corrosion test 3 h at 100°C max.	1	1	1	—	—	—	—	BS 2000: Part 154	IP 154
Cold filter plugging point, °C, max. Summer (16 March to 30 September inclusive)	—	—	−4	—	—	—	—	BS 6188: (EN 16)	IP 309/80
Winter (1 October to 15 March inclusive)	—	—	−12	—	—	—	—		
Smoke point, mm, min.	35	20	—	—	—	—	—	BS 2000: Part 57	IP 57
Char value, mg/kg, max.	10	20	—	—	—	—	—	BS 2000: Part 10	IP 10

BS 2869 (Ref. 9)

1 cSt = 1 mm²/s.

*IP 53/82 superseded IP 53/70 which was technically equivalent to BS 4382. IP 53/70 only specified the expression of sediment as percentage by mass while IP 53/82 specifies expression of sediment as either percentage by mass or percentage by volume but is otherwise still technically equivalent.

†This limit is set in accordance with the legislative requirements for gas-oil of the "Council Directive (75/716/EEC of the European Economic Community) on the approximation of the laws of the Member States relating to the sulphur content of certain liquid fuels" as embodied in Statutory Instrument 1976 No. 1988.

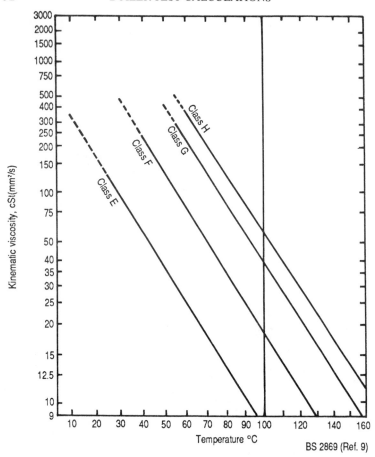

Fig. 3.1 Viscosity–temperature chart for fuel oils

family 39.2–55.0 MJ/m³, and the third family 73.4–87.6 MJ/m³. The first and second families are further divided into groups, on the basis that all Wobbe numbers in any one group are those of gases which would be fully interchangeable when used on a given appliance. Examples of how well-known gases are placed in the family classification are as follows: town (coal) gas is a first family gas, natural gas is in the second family, and LPG is in the third family.

3.5 Calorific value

When any fuel containing hydrogen is burned, the hydrogen combines with oxygen to form water, which, together with any other moisture present in the fuel, appears as water vapour in the products of combustion. The gross calorific value of the fuel includes the latent heat of evaporation of this vapour; in other words, it declares the total amount of heat which would be recovered if the products of combustion were cooled to the datum temperature used for the determination of calorific values (25°C for solid and liquid fuels, 15°C for gaseous fuels).

The net calorific value excludes the latent heat of evaporation and is therefore always lower than the gross value. Note however that, as stated, the terms have significance only in relation to fuels containing hydrogen or water. Pure carbon, for example, has only one calorific value.

Both gross and net calorific values may be declared on a constant-pressure or constant-volume basis. In most boilers combustion takes place at or very close to atmospheric pressure and is regarded as a constant-pressure process, but the calorific values of solid and liquid fuels are determined using a bomb calorimeter, in which combustion takes place at constant volume. In the case of gaseous fuels the calorific value is determined at constant pressure. The full definitions of the four calorific values are given in Ref. 11, together with formulae for converting from one basis to another. The formulae of principal interest are reproduced in Appendix 5.

The more detailed published test procedures normally specify the constant-pressure values for use in calculations, but in the case of solid fuel the necessary correction to the constant-volume value is very small and may reasonably be neglected.

A choice has to be made between gross and net calorific values as the basis for calculating the heat input to a unit under test. It is a matter which used to provoke considerable argument, especially between the coal and oil interests. Thus, the "oil" argument would be that, as a boiler could not in practice condense the water vapour in the flue gases without causing intense acidic corrosion due to the water combining with the oxides of sulphur produced during combustion, it was unrealistic to judge the performance of a boiler against a standard which was unattainable.

Of course, with oil being richer in hydrogen than coal, then for the same heat output there is proportionately more heat lost in water vapour in the flue gas, and if this heat is discounted by using net calorific values as the basis for calculation for heat input, the apparent increase in thermal efficiency is greater for oil than for coal.

This naturally offended the "coal" lobby, which would respond by claiming that if it was fair to deduct the heat in water vapour from the heat input to the boiler, it was equally fair to deduct the carbon in riddlings loss or the carbon in ash loss in the case of solid fuel units, because, in similar fashion, the unit had never been given the chance to make use of the heat rejected in these ways.

In fact, the advantages did not all lie one way. Certainly, with thermal efficiencies being quoted in promotional literature as important selling points, the use of net calorific values favoured hydrogen-rich fuels. On the other hand, if a monetary relationship to net calorific value such as pence/GJ is quoted, the advantage goes the other way, in favour of the low-hydrogen fuel.

The wrangling over this matter has become less meaningful in recent years due to the greatly increased use of natural gas as a fuel and the introduction of units with condensing recuperators. The use of this clean sulphur-free fuel enables such units to be operated with exceptionally low flue-gas outlet temperatures and with considerable condensation of the flue-gas moisture. Such units can be tested only on the basis of the gross calorific value — on the net basis they could have a thermal efficiency in excess of 100% — so the practical utilization of gross calorific values is not now seen as something theoretically impossible.

In general, as long as the calorific value selected, whether gross or net, is clearly stated and the implications clearly understood, choice is merely a matter of following site practice or for agreement between the parties to the test.

4. Formulae used in preparing a heat account

4.1 Introduction

In this section, formulae typical of those used in the preparation of a heat account are presented, together with explanatory comments and, in some instances, worked examples. The derivations of more complicated test and related formulae are explained separately in the Appendices. The formulae are "typical" because they are not entirely identical to those in any particular test code. This is not important because although allowance must always be made for the broader policy differences between test codes which stem largely from their different ways of defining the boundary of the unit under test, there are recognizable similarities in the construction of individual formulae. The differences which do exist in relation to the more important formulae, if considered significant, are discussed in detail in the text.

Note that in many cases the data presented and the results obtained from worked examples are shown with an unrealistic number of significant figures. This has been done to reduce the cumulative errors that result from repeated "rounding off" in the stages of lengthy calculations, and thereby achieve more precise correlation between overall results which ought to be identical in hypothetical calculations.

4.2 Heat input

One of the basic requirements for preparing a heat account is a precise knowledge of the amount of heat entering the unit under test, because the thermal efficiency and various losses are all finally expressed as proportions or percentages of that heat input.

In some test codes the heat input as used in calculations is expressed as power, that is, as the average rate of heat input over the test period. In other cases the heat input is based upon the unit mass of fuel fired, and therefore if there are no extra supplies of heat to the unit, the heat input is the calorific value of unit mass of the fuel. This latter approach is developed further when formulae of the Siegert type (see 4.4.2) are used to calculate the losses due to sensible heat and unburnt carbon monoxide in the dry flue gas, because the heat input does not apparently feature in such calculations at all. In fact, its presence is disguised because the ratio of carbon content to calorific value is treated as a constant for each type of fuel and therefore the calorific value forms part of the overall constant which appears as a factor in such formulae.

It is often the case that apart from the heat flows into a unit resulting from the calorific value of the fuel, a number of other heat flows which apparently contribute to the total heat input will be identified. Whether or not they actually are included in the heat input depends entirely upon where the unit boundary is drawn, because if the route which they take lies entirely within the boundary, they may be treated as closed loops (see 2.2) and not included in the heat input.

It is usually in larger and more complicated units that the heat input will comprise a number of heat flows, and whilst it would be impracticable to attempt to allow for every possible arrangement that might be devised, the more familiar of the extra heat flows are examined in this Section. Note that only Refs. 4, 5 and 7 give consideration to the heat input comprising such a wide range of heat flows as that which follows. Refs. 1, 2 and 6 do not specifically investigate minor heat flows other than H12 for liquid fuels. Ref. 3 also deals with H12 but is intended for units which are thermodynamically simple, that is, having a single major source of heat input and a simple circuit for the heat transfer medium.

4.2.1   *Heat input from calorific value of fuel*

The heat input from the calorific value of the fuel is based upon either the gross or net calorific value, and whichever basis is chosen must be stated in the test results. To permit the use of the same formulae for all fuels, gaseous fuel characteristics are assumed to have been converted to a mass basis (see Appendix 6).

For tests based on the gross calorific value,

$$H11 = B11 \times Q11$$

For tests based on the net calorific value,

$$H11 = B11 \times Q11N$$

In the case of pulverized fuel firing, if
(a)   the milling takes place on site,
(b)   the fuel is weighed prior to milling,
(c)   the mill is considered to lie outside the unit boundary (see 2.3.10), then

$$H11 = W11 \times Q11 - \left(\frac{R65 \times C65 \times Q65}{100} \right)$$

or

$$H11 = W11 \times Q11N - \left(\frac{R65 \times C65 \times Q65}{100} \right)$$

From the data provided in Appendix 3 for the unit considered as an example and which is fired by a mechanical stoker,

$$H11 = 1.3 \times 27\,800$$
$$= 36\,974 \text{ kW}$$

4.2.2 *Heat input from sensible heat in fuel oil*

This heat input results from the addition of sensible heat to the heavier fuel oils to reduce their viscosity for the purposes of pumping and atomization. Two situations may be encountered — one in which all of the sensible heat is provided from an external source so that the fuel enters the unit at high temperature and $T11 = T12$, or alternatively, one in which some "secondary" heating is provided within the unit itself, raising temperature $T11$ at entry to the unit, to $T12$ at entry to the furnace. Section 2.3.9 refers to alternative treatments in various test codes, but one major variation arises as a result either of treating the whole of the sensible heat in the oil over temperature $T11$ at heat input, or of treating as heat input only that portion of it which is not supplied by the unit itself. The calculation of heat quantities in either case is as follows:

Sensible heat in fuel entering unit

$$H12 = B11 \times (T11 - T1) \times S12$$

The essential difference between the two approaches stems from the position of the unit boundary, because if this is drawn after the secondary heater, $T11$ will be much higher than if the boundary were drawn before the heater.

The mean specific heat $S12$ for the fuel oil is obtained from appropriate sources of reference.

4.2.3 *Heat input from sensible heat in gaseous fuel*

Gas distributed through public supply mains is usually treated as being at datum temperature, but when a gaseous fuel is generated on-site this may be at an elevated temperature $T12$ and may therefore contribute some sensible heat input, that is

$$H12 = B11 \times (T12 - T1) \times S12$$

The mean specific heat $S12$ for the gas under consideration is obtained from appropriate sources of reference or may be calculated following the method given in Appendix 11 for flue gas.

4.2.4 *Heat input from sensible heat in solid fuel*

Solid fuel is usually treated as being at datum temperature and therefore as making no sensible heat contribution to heat input. In the case of pulverized fuel firing, and especially in direct systems where the milled coal passes directly from the mill to the furnace, the coal will be at an elevated temperature due partly to the work done upon it, but mainly through contact with the heated air or gas supplied to the mill for drying purposes.

The sensible heat in the coal may be calculated in the usual way from its mass flow, temperature, and specific heat, but it is simpler in most cases to treat the coal as being at datum temperature and to consider only the extra heat supplied along with it from other sources. However, in boiler practice the drying medium is usually hot air or flue gas drawn from the unit itself, so that although in Refs. 4 and 5 the mill is outside the unit boundary, the flow of air or flue gas for drying may be regarded as a closed loop and therefore disregarded in the heat account. If the drying medium is supplied to the mill from an entirely separate source, appropriate measurements for obtaining the heat input must be taken.

In Refs. 4 and 5, the shaft power to the mill is not included in the energy input to the unit (see 2.3.10) but this is perhaps dubious because no allowance is thereby made for any heat input from this source. In Ref. 7, where the mill lies inside the unit boundary, the power supplied to it is included in the heat flow credited to the input.

4.2.5   *Heat input from steam or water discharged into furnace*
Steam may be discharged into a boiler furnace for various reasons, including such purposes as atomizing fuel oil, cooling mechanical stoker grate bars, or creating turbulence in the combustion zone. Water may be used instead of steam as a coolant. If supplied from a separate source, this heat flow represents an additional heat input above datum temperature. If supplied by the unit itself it is treated in Ref. 4 as comprising a closed loop and not counted as contributing to either the heat input or output, whereas in Ref. 7 the flow of heat is regarded as part of both the input and output of the unit. In either case, the quantity of steam or water must be measured in order to determine the contribution it makes to the loss in the flue gas due to heat in the moisture. When the heat in the steam or water is treated as part of the input to the unit,

$$H13 = W13 \times (E13 - E10)$$

4.2.6   *Heat input from air from a separate source*
If atomizing or combustion air is preheated and receives its heat from the unit itself, this is treated in Ref. 4 as a closed loop and ignored. If the heat comes from an external source it is included in the heat input in Ref. 4. In Ref. 5 this heat is deducted from the output, whilst in Ref. 7 it is credited in all circumstances to the input:

$$H14 = W14 \times E14 - E1)$$

4.2.7   *Heat input from substances injected into furnace*
Various substances may be added to a boiler furnace whilst it is operating. The addition of the inert material comprising a fluidized bed is one example. Chemical additives may be injected into the furnace for various claimed reasons, but usually a principal objective is to control deposition on heat transfer surfaces and so extend the time between cleaning operations, thereby allowing the unit to be on-line for longer periods. Other substances, such as limestone, may be introduced to moderate the emission of atmospheric pollutants.

During a test it may be preferred to suspend the injection of all these substances, but if this is impracticable or undesirable, it should be ascertained whether their use involves an increase or decrease of the heat released in the furnace. It should be noted that a further consideration which may follow from their introduction is how the composition of the flue gas may be affected. Ref. 4 makes allowance for this heat flow:

$$H15 = \pm W15 \times Q15$$

4.2.8   *Heat input from mechanical energy of ancillaries*

As discussed in 2.3.3, the mechanical energy of ancillary equipment such as fans and pumps may possibly make a significant contribution to the heat input to a unit. When the unit boundary is established, the machines included within it will have their electrical or other motive power measured during the test, and the shaft power calculated, or determined from manufacturers' characteristic curves. A possible exception to this treatment is the mechanical energy supplied to coal pulverizing mills (see 2.3.10); even where the mills are considered to lie outside the unit boundary, to include the mechanical energy is the simplest method of allowing for this source of heat input. The sum of all the shaft powers comprises H16. Should preliminary estimation suggest that H16 would amount to no more than 0.25% of the total input, Ref. 4 suggests that it may be ignored in the direct method of test.

4.2.9   *Heat input from waste heat*

A waste heat boiler receives heat from the sensible heat in a flow of gas:

$$H17 = B11 \times (E17 - E18)$$

4.2.10   *Heat input from electricity*

In an electrical boiler, the input to the unit, H18, is directly measured.

4.2.11   *Heat input (total) to unit*

The total heat input H1 comprises the sum of the individual heat inputs, these being identified by the test code being followed. Table 4.1 indicates the guidance which test codes provide on this point.

From the data provided in Appendix 3 for the unit considered as an example,

$$H1 = H11$$
$$= 36\ 974\ kW$$

4.3   Heat output from unit

The total heat output H35 from the unit is given by the sum of the heat flows in the steam/water flows from the unit less the sum of the heat flows in the feed and other water flows into the unit. As this total heat is

Table 4.1 Heat flows comprising heat input and output in test codes

Symbol(s)	Source of heat	Heat flows included/not included in heat input — Reference No.							Heat flows included/not included in heat output — Reference No.						
		1	2	3	4	5	6	7	1	2	3	4	5	6	7
H11	Calorific value of the fuel	a	a	a	a	a	a	a							
H12	Sensible heat in the fuel	f	f	f	a	a	f	a					f		
H13	Steam or water discharged into the furnace				a			a							
H14	Air supplied from a separate source				a			a					b		
H15	Substances injected into the furnace				c										
H16	Mechanical energy of ancillaries					a		a					d		
H17	Sensible heat in waste gas at inlet to the unit				a			a							
H18	Electricity (as "fuel")				a										
H21, 22, 23, 26, 28, 29	Principal steam and/or water flows					a		a	a	a	a	a	a	a	a
H25	Blowdown	e	e		b			a				b			a
H27	Steam returned to unit for use on ancillaries	e	e		b			a				b			a

Key

a Included   d Added only in the case of indirect tests
b Not included   e Treatment not clearly defined
c Positive or negative   f Reference only to fuel oil

*The more detailed test codes such as Refs. 4 and 7 include a few other heat flows as well as those listed here.

obtained by difference, the datum temperature of the individual heat flow enthalpies may be considered unimportant, and the values above 0°C in standard tables used. This procedure has been adopted in Refs. 1, 2, 3, 5 and 7.

However, in the case of the heat flow in blowdown, this may be credited to unit output for contractual commissioning tests but treated as a loss in routine tests (see 2.3.8) and this heat flow is reckoned above datum temperature. To enable the heat in blowdown to be calculated on the same basis for either situation it may be helpful to calculate all the heat flows in the heated medium, steam or water, to be above the enthalpy of water at the datum temperature for the test, that is, at temperature T1. This procedure has been followed here and also in Ref. 4, although that document does not specifically make provision for alternative treatment of the heat in blowdown.

Note that only Refs. 4, 5 and 7 give consideration to the heat output comprising such a wide range of heat flows as that which follows. Refs. 1, 2 and 6 do not specifically investigate minor heat flows, whilst Ref. 3 is intended for use with units which are thermodynamically simple, that is, having a single major source of heat input and a simple circuit for the heat transfer medium.

Calculation of the individual quantities comprising the heat output is straightforward and it has not been considered necessary to provide worked examples.

4.3.1   *Heat flow in feed or return water*

$$H21 = W21 \times (E21 - E10)$$

4.3.2   *Heat flow in main steam attemperator water*

$$H22 = W22 \times (E22 - E10)$$

4.3.3   *Heat flow in reheated steam attemperator water*

$$H23 = W23 \times (E23 - E10)$$

4.3.4   *Heat flow due to variation in water-level*

The amount of water required to raise or lower the drum water-level is ascertained prior to the test. The change in level over the test period and the density of water at the temperature of the drum water (corresponding to the saturation pressure of the drum) are then used to determine the apparent flow $\pm W24$ causing the change of level.

For an increase in level

$$H24 = + W24 \times (E25 - E10)$$

For a decrease in level

$$H24 = - W24 \times (E25 - E10)$$

4.3.5   *Heat flow in blowdown*
See 2.3.8 and 4.3.

$$H25 = W25 \times (E25 - E10)$$

4.3.6   *Heat flow in main steam or water*

$$H26 = W26 \times (E26 - E10)$$

4.3.7   *Heat flow in steam returned to unit for use on ancillaries*

$$H27 = W27 \times (E27 - E10)$$

In Ref. 4, this heat flow is treated as a closed loop (see 2.2 and 2.3.3) and is not therefore included in the heat flows comprising H35. When Ref. 7 is followed, this heat is simply part of the output, or if measured separately, it will be credited to the output. In either case the mass flow of the steam, W27, is required if one of the other steam or water flows into or out of the unit cannot be measured directly and has to be calculated by difference from the other measured quantities of steam and water.

4.3.8   *Heat flow in steam returned to unit for reheating*

$$H28 = W28 \times (E28 - E10)$$

4.3.9   *Heat flow in reheated steam*

$$H29 = W29 \times (E29 - E10)$$

4.3.10   *Heat flow in water at entry to condensing recuperator*
This applies only when the water at entry to the recuperator is separate from the feed water.

$$H31 = W31 \times (E31 - E10)$$

4.3.11   *Heat flow in water at outlet of condensing recuperator*
This applies only when the water at entry to the recuperator is separate from the feed water.

$$H32 = W31 \times (E32 - E10)$$

4.3.12   *Heat output*
As stated in 4.3, the total heat output H35 comprises the sum of the individual heat outputs in steam or water flowing from the unit, less the sum of the individual heat flows in steam or water entering the unit. All test codes are in agreement with this insofar as it applies to the principal flows of steam and/or water. Differences do occur, however, in relation to minor heat flows, depending upon where the unit boundary is deemed to lie. Table 4.1 indicates the guidance provided in the test codes listed as references.

In the data provided in Appendix 3 for the unit considered as an example it is assumed that H35 comprises 30 711.381 kW.

4.4   Heat losses

In a direct test the heat losses should be obtained in order to build up an informative heat account, but in an indirect test they must be obtained to enable the output or input of the unit to be calculated. In Ref. 4 the losses are first calculated in terms of heat flows (kW), that is, on the same basis as the heat input and output (although the final heat account is in terms of percentages). In Ref. 5 the losses are first calculated in terms of kJ/kg of fuel and, in Ref. 7, in Btu/lb fuel. Refs. 1, 2, 3 and 6 declare them directly as percentages and this practice will be followed here. Note, however, that when comparisons between test code formulae are being made, it may only be possible to state that they are similar rather than identical, because of these differences in approach and units. Where worked examples are provided, these are all based upon the data provided in Appendix 3.

4.4.1   *Loss due to heat in blowdown*
As stated in 2.3.8, if blowdown takes place at all, the heat in it is not counted as a loss in a contractual boiler test, and it is for that type of test that test codes are primarily written. In assessing the performance of the boiler in its ordinary working environment, the heat in blowdown, if not recuperated, is regarded as a loss:

$$L25 = \frac{H25 \times 100}{H1}$$

4.4.2.   *Loss due to sensible heat in dry flue gas*
This is normally the largest and therefore the most important of all the losses. Appendix 12 provides the derivation of the various formulae used and shows that the general expression for the sensible heat loss in the dry flue gas is

$$L41 = \frac{100 \times \left(MC - MCR + \dfrac{MS}{2.67}\right) \times S41 \times (T2 - T1) \times B11 \times 100}{12.011 \times (VCO2 + VCO + \Sigma VCXHY) \times H1}$$

The factor MS/2.67 is included only when absorptive-type flue-gas analysers are used in which VCO2 and VSO2 are combined in a single reading.

Similar formulae are to be found in Refs. 4, 5 and 7, although in Ref. 7 the calculation is carried out in several separate stages. An abbreviated formula of the Siegert type is used in the more simplified test codes Refs. 1, 2, 3 and 6, but in Refs. 1 and 2 it is recommended that the fundamental procedure in Ref. 5 be used in the case of gaseous fuels. In Ref. 6, special formulae are provided for gaseous fuels which combine flue-gas losses

L41 and L51 (or L52) into one. Taking for the purpose of illustration the data provided in Appendix 3, it is informative to compare the results obtained from the formulae in the following circumstances.

(a) Fundamental, with a true value for VCO2:

$$L41 = \frac{100 \times (MC - MCR) \times S41 \times (T2 - T1) \times 100}{12.011 \times (VCO2 + VCO) \times Q11}$$

$$= \frac{100 \times (0.678 - 0.0244) \times 30.6 \times 130 \times 100}{12.011 \times (12.19 + 0.46) \times 27\,800}$$

$$= 6.1555\%$$

(b) Fundamental, ignoring VCO:

$$L41 = \frac{100 \times (0.678 - 0.0244) \times 30.6 \times 130 \times 100}{12.011 \times 12.19 \times 27\,800}$$

$$= 6.3878\%$$

(c) Siegert, using a specially calculated constant K41 and with the factor 0.012 as used in Ref. 1:

$$L41 = \frac{0.6213 \times 130 \times [1 - (0.012 \times 2.9706)]}{12.19}$$

$$= 6.3896\%$$

(d) Siegert, in the form used in Ref. 3:

$$L41 = \frac{0.62 \times 130 \times [1 - (0.01 \times 2.9706)]}{12.19}$$

$$= 6.4156\%$$

It is clear that when the fundamental and Siegert formulae are compared on equal terms, that is, when a true reading of VCO2 is obtained (without the addition of VSO2), when carbon monoxide is not present or is ignored, and when a Siegert constant is specially derived for the fuel in use, both give very similar results, as in (b) and (c). Where these conditions do not apply, as is the case when realistic test data are used in a fundamental formula as in (a) and compared with the use of Siegert formulae as in (c) or (d), then errors are inevitable. The effect on the overall thermal efficiency is small, but in the case of (c) this would be reduced by 6.3896−6.1555, or 0.23%, whilst in the case of (d) it would be reduced by 6.4156−6.1555, or 0.26%.

The use of the fundamental formula eliminates unnecessary doubts as to the accuracy of the calculation, and this is of course reflected in current test codes where the more detailed methods always employ the fundamental formula. However, there is no doubt that the Siegert formula will continue to be widely used for simple, quick determinations of the heat loss in the dry flue gas.

4.4.3   Loss due to unburnt carbon monoxide in flue gas

Although this loss may sometimes be ignored in the simplest tests, it may not in fact be insignificant, as the example given subsequently will show. Appendix 13 provides the derivation of the full expression for the loss due to carbon monoxide, thus:

$$L42 = \frac{28.011 \times VCO \times \left(MC - MCR + \dfrac{MS}{2.67}\right) \times B11 \times Q42 \times 100}{12.011 \times (VCO2 + VCO + \Sigma CXHY) \times H1}$$

The factor $MS/2.67$ is included only when absorptive-type flue-gas analysers are in which VCO2 and VSO2 are combined in a single reading.

A similar formula is to be found in Refs. 4 and 5. Ref. 7 adopts a different approach in which the heat loss is calculated from the quantity of partially burnt carbon and not the quantity of carbon monoxide (see Appendix 13). Again, an abbreviated formula of the Siegert type is used in simpler test codes such as Refs. 1, 2, 3 and 6, but in Refs. 1 and 2 it is recommended that the more fundamental procedure in Ref. 5 should be used in the case of gaseous fuels. In Ref. 6 unburnt carbon monoxide is not considered in the case of gaseous fuels. Taking for the purpose of illustration the data provided in Appendix 3,

$$L42 = \frac{28.011 \times 0.46 \times (0.678 - 0.0244) \times 1.33 \times 10\ 100 \times 100}{12.011 \times (12.19 + 0.46) \times 1.33 \times 27\ 800}$$

$$= 2.0137\%$$

It is clear that a large loss results from a comparatively small percentage of carbon monoxide in the flue gas. It follows therefore that if a test is to be effective, the flue gas should be checked for the presence of this gas.

In Ref. 3, a formula of the Siegert type is given for L42, as follows:

$$L42 = \frac{K42 \times VCO[1 - (0.01 \times L67)]}{(VCO2 + VCO)}$$

The constant K42 is given as 63 for bituminous coal; thus in the example,

$$L42 = \frac{62 \times 0.46[1 - (0.01 \times 2.9706)]}{(12.19 + 0.46)}$$

$$= 2.2229\%$$

As shown in Appendix 13, K42 may be calculated and for the fuel actually used in the example K42 = 57.4. Using this value and the multiplying factor (0.012) used in Ref. 1,

$$L42 = \frac{57.4 \times 0.46[1 - (0.012 \times 2.9706)]}{(12.19 + 0.46)}$$

$$= 2.0129\%$$

The same arguments apply when comparing the two types of formulae for L42 as for L41. When compared on equal terms, that is, with true values of VCO2 obtained (without the addition of VSO2) and when a constant is specially calculated for the fuel in question, both fundamental and simplified expressions give similar results. However, if these conditions do not apply, the fundamental formula eliminates unnecessary doubts concerning the accuracy of the calculations. Nonetheless the Siegert type formula gives reasonably good results for simple, quick determinations of the heat loss due to unburnt carbon monoxide.

4.4.4 *Loss due to unburnt hydrocarbon gas in flue gas*

Combustible hydrocarbon gases designated by the general symbol CXHY may be found in flue gas, but it is generally considered that only the merest traces will be found, except when combustion is obviously bad, that is, when smoke is being formed. Usually, the flue gas is not even monitored for CXHY gases, but if their presence is noted and measured, each one is treated in similar fashion to carbon monoxide (see 4.4.3).

$$L43 = \frac{(12.011X + 1.008Y) \times VCXHY \times \times \left(MC - MCR + \dfrac{MS}{2.67} \right) \times B11 \times Q43 \times 100}{12.011 \times (VCO2 + VCO + \Sigma VCXHY) \times H1}$$

The factor MS/2.67 is included only if absorptive-type flue-gas analysers are used in which VCO2 and VSO2 are combined in a single reading.

A similar formula is to be found in Refs. 4 and 5. Ref. 7 adopts a different approach (see Appendix 13), and also provides a method for calculating the loss due to unburnt hydrogen in the flue gas. Refs. 1, 2, 3 and 6 do not consider any unburnt gases other than carbon monoxide.

4.4.5 *Loss due to heat in gas discharged from waste heat units*

In the case of waste heat units, combustion is not involved and the composition of the gas discharged from the unit is the same as that at entry:

$$L44 = \frac{B11 \times (E18 - E19) \times 100}{H1}$$

4.4.6 *Loss due to heat in water vapour in flue gas due to moisture from fuel*

The water vapour in the flue gas due to moisture from the fuel comes from two sources:

(i) from water actually present in the fuel,
(ii) from the formation of water due to combustion of hydrogen in the fuel.

In the flue gas, the moisture is normally present in the form of super-heated steam, so it contains

(i) sensible heat between the ambient temperature $T1$ and the saturation temperature $T51$,
(ii) latent heat of evaporation at temperature $T51$ and pressure $P51$,
(iii) sensible heat due to superheat between the saturation temperature $T51$ and the flue gas temperature $T2$,

(a) loss due to sensible and latent heat in moisture from the fuel

This loss applies only to tests based on the gross calorific value of the fuel because it is only that calorific value which includes the latent heat in any water present.

$$L51 = \frac{B11 \times (MH20 + 8.936MH2) \times \times [S51(T51 - T1) + J51 + S52(T2 - T51)] \times 100}{H1}$$

This is a general formula and is the basis for determining $L51$ in Refs. 1, 2, 3, 4 and 6, except that the factor 8.936 (the mass of water produced from combustion of unit mass of hydrogen) is rounded to 9. Some, too, assume values for $S51$, $T51$, $J51$ and $S52$ and then present the expression in an abbreviated form. Ref. 7 does not consider the sensible heat, latent heat and superheat quantities as separate factors but expresses the heat content over datum as the difference between the enthalpies of vapour at $P51$ and $T51$, and of water at $T1$ or the chosen datum temperature.

Test codes differ in the guidance they provide on the saturation pressure of the water vapour (see Table A3.3). Some simply assume that $T51$ is atmospheric pressure, although the vapour must actually be at a partial pressure appropriate to the proportion of the total volume of the flue gas which it occupies, and some test codes suggest a value for this. A method of calculating this partial pressure is given in Appendix 14.

However, taking as an example the data in Table A3.2, it is interesting to compare the values of $L51$ obtained when the saturation temperature $T51$ is appropriate to a partial pressure of 70 mbar (as in Ref. 4) in the one case, and atmospheric pressure in the other.

Thus when $P51$ is 70 mbar (absolute),
(Published values give $T51$ as 39°C and $J51$ as 2409.2 kJ/kg)

$$L51 = \frac{1.33 \times [0.11 + (8.936 \times 0.042)] \times \times [4.2(39 - 20) + 2409.2 + 1.88(150 - 39)] \times 100}{1.33 \times 27\,800}$$

$$= 4.7094\%$$

When P51 is atmospheric pressure
(Published values give T51 as 100°C and J51 at 2257 kJ/kg)

$$L51 = \frac{\begin{array}{c}[0.11 \times (8.936 \times 0.042)] \times \\ \times [4.2(100-20) + 2257 + 1.88(150-100)] \times 100\end{array}}{27\,800}$$

$$= 4.6907\%.$$

Comparing the two values, it is clear that within broad limits the choice of saturation pressure P51 does not significantly affect the result.

Refs. 2, 3 and 6 include abbreviated formulae in metric units in which the heat input is taken as the calorific value of unit mass of the fuel. P51 is taken as atmospheric pressure, T51 as 100°C, S51 as 4.2 kJ/(kg°C) and S52 as 2.1 kJ/(kg°C). Depending on source, J51 may be taken as 2257 kJ/kg at atmospheric pressure, so that when the equation has these values inserted it should reduce to

$$L51 = \frac{(MH20 + 9MH2) \times (2467 - 4.2\,T1 + 2.1\,T1 \times 100}{Q11}$$

Refs. 2 and 6 give the first term in the second brackets as 2460, whilst Ref. 3 gives it a value of 2488.

Ref. 5 recommends that T51 be taken to be 25°C on the grounds that this is the datum temperature for establishing the calorific values of solid and liquid fuels, but the reasoning behind this is not apparent.

(b) Loss due to sensible heat in moisture from fuel

This loss applies only to tests based on the net calorific value of the fuel because it is only that calorific value which excludes the latent heat of evaporation of any moisture from the fuel.

$$L52 = \frac{\begin{array}{c}B11 \times (MH20 + 8.936\,MH2) \times \\ \times [S51(T51 - T1) + S52(T2 - T51)] \times 100\end{array}}{H1}$$

This is a general formula and is the basis for determining L52 in Refs. 1, 2, 3, 4 and 6, but with the factor 8.936 rounded to 9. Some, too, assume values for S51, T51 and S52 and then present the expression in abbreviated form. Ref. 7 does not consider tests based upon the net calorific value.

In Refs. 2, 3 and 6, S51 is taken as 4.2 kJ/(kg°C), S52 as 2.1 kJ/(kg°C), T51 as 100°C, and with the loss based upon the net calorific value of unit mass of the fuel, the expression is then abbreviated to

$$L52 = \frac{(MH20 + 9MH2) \times [210 - 4.2T1 + 2.1T2] \times 100}{Q11N}$$

4.4.7 *Loss due to heat in steam or water discharged into furnace*

Steam is injected into boiler furnaces for such purposes as atomizing oil fuel, cooling grate bars, and promoting turbulence to aid combustion. Water may be used instead of steam as a coolant. The steam in the flue gas is treated in the same way as the water vapour resulting from the fuel, except that the latent heat is included for tests carried out on both the gross and net bases. Thus

$$L53 = \frac{W13 \times [S51(T51 - T1) + J51 + S52(T2 - T51)] \times 100}{H1}$$

A formula similar to this appears in Ref. 4. Ref. 7 allows for this loss but expresses the heat content of unit mass of the steam in terms of enthalpy values. Ref. 3 suggests that this loss be calculated in the same way as that due to moisture in the combustion air; that loss, however, is based upon a different concept in which only the loss of superheat is involved (see 4.4.8).

4.4.8 *Loss due to heat in moisture in combustion air*

The moisture in combustion air due to humidity is originally super-heated, and it is assumed that if the flue gas were to be cooled to ambient temperature $T1$ the same quantity of moisture would again be retained. The heat lost therefore, is due to the fall in the extra superheat gained by contact with the flue gas and represented by $(T2 - T1)°C$.

The quantity of vapour present due to humidity is obtained from a psychrometric chart, being read off directly in terms of mass/unit mass of air $(K54)$ when dry-bulb temperature $T1$ and wet-bulb temperature $T3$ are known. The loss due to moisture in the mass of combustion air $W1$ is then

$$L54 = \frac{W1 \times K54 \times S52 \times (T2 - T1) \times 100}{H1}$$

Expressions of this type appear in Refs. 3, 4 and 5. Ref. 7 substitutes enthalpy values in place of the specific heat and temperatures.

Methods for the determination of $W1$ from test results are given in Appendices 17 and 18. The specific heat of superheated steam $S52$, is as indicated in Table A3.3.

As an example of the order of loss represented by L54 and again taking the data given in Appendix 3 together with the air quantity derived in Appendix 9,

Air for combustion $W1 = 0.4370 \times 28.97$

$$= 12.66 \text{ kg/kg fuel}$$

where 28.97 is the equivalent relative molecular mass of air.

Assuming that K54 from a psychrometric chart is 0.01 kg water/kg air, then

$$L54 = \frac{12.67 \times 0.01 \times 1.88 \times (150 - 20) \times 100}{36\,974}$$

$$= 0.11\%$$

This is insignificant and has been disregarded in the worked example. It serves to show that only exceptionally large quantities of excess air, high flue-gas temperatures and/or high humidities would give values for L54 large enough to be significant.

4.4.9   *Loss due to heat in condensed flue-gas moisture*

In a unit of the condensing type, the flue gas is cooled to a comparatively low temperature, perhaps 30–35°C, resulting in flue-gas moisture being discharged in the liquid state. This makes it possible to achieve an exceptionally high thermal efficiency, but obviously suitable arrangements have to be made to accommodate a low feed-return temperature or to provide a separate water circuit to cool the flue gas.

Special arrangements must also be made to accommodate the acidic products which form at such low temperatures. Even an exceptionally clean fuel such as North Sea natural gas will fix atmospheric nitrogen into acidic compounds, so the low-temperature boiler surfaces must be capable of resisting acid attack.

The loss due to condensed moisture

$$L55 = \frac{W55 \times S51 \times (T55 - T1) \times 100}{H1}$$

An expression similar to this appears in Ref. 4.

Obviously, L55 can only apply to tests carried out on the basis of the gross calorific value, because then the latent heat of evaporation is recoverable. Were the calculation to be based on net calorific value, the unit efficiency could exceed 100%.

4.4.10   *Loss due to heat in residual moisture in flue gas*

The "residual moisture" is that which remains in the flue gas when other moisture has been condensed, so again this loss applies only to condensing units and tests on the basis of the gross calorific value.

$$L56 = \frac{W56 \times [S51(T51 - T1) + J51 + S52(T2 - T51)] \times 100}{H1}$$

An expression similar to this appears in Ref. 4.

Note that the amount of water vapour is normally

$$W56 = [B11 \times (MH20 + 8.936\,MH2)] - W55$$

but if other quantities of moisture such as W13 were originally present, they must be added to the quantity represented by $[B11 \times (MH20 + 8.936 \, MH2)]$.

4.4.11 *Loss due to heat in evaporated ash cooling-water in flue gas*

In some units, evaporated water from ash-quenching tanks or conveying sluices passes out with the flue gas, giving rise to a further heat in moisture loss.

$$L57 = \frac{W57 \times [S51(T51 - T1) + J51 + S52(T2 - T51)] \times 100}{H1}$$

An expression similar to this appears in Ref. 4.

Note that W57, the amount of water evaporated,

$$= W78 - W79$$

4.4.12 *Losses due to heat in unburnt combustible matter in solid residues*

Various forms of solid residue result from the burning of solid fuel, and also, to a much more limited extent, from liquid and gaseous fuels.

In a solid fuel boiler where combustion takes place on a grate there may be riddlings of raw fuel at or near the entry to the grate. These are particles of fuel which trickle directly between the grate bars and do not therefore get carried forward into the furnace for combustion. Arrangements may be made to collect riddlings and reintroduce them into the fuel supply or furnace.

In the boiler furnace solid ash and/or clinker (fused ash) are usually produced as a result of combustion on a grate or by a pulverized fuel burner. In a slagging combustor, the ash is discharged in molten form.

Grit and dust are produced and these are conveyed by the flue gases to parts of the boiler downstream from the furnace. They may tend to build up in the furnace tubes, reversal chambers and smoke tubes of shell boilers, and in both shell and water-tube boilers it is usual to install hoppers or cyclone arrestors with hoppers to trap such grit. Often too, arrangements are made to recycle the grit back to the furnace.

Fine dust is also produced, which when present in appreciable quantities, as when coal is burnt in pulverized form, will require the fitting of high-efficiency dust arrestment plant for effective trapping. Even so, some small proportion will inevitably be present in the flue gas emitted from the chimney.

A type of solid residue produced by fluidized bed combustion equipment comprises a mixture of ash and the sand or other inert material which forms the bed. Even when equipment is installed for recycling bed material back to the furnace, some proportion of it will still be present in the residue discharged as waste.

The foregoing references to grit and dust apply also to the heavier ash-bearing oil fuels. Gaseous fuels as supplied through public distribution mains may usually be considered as almost entirely dust free, but fuel gases produced on-site by gas producers, or from certain processes such as steel production, may contain appreciable dust quantities. Waste heat boilers in which only sensible heat is extracted from hot gas produced in some process may also receive gas with a significant dust burden.

In the case of solid or liquid fuel boilers, all solid residues contain combustible matter. This is usually assumed to be carbon, with the appropriate calorific value, unless tests are carried out and a specific calorific value obtained. Ref. 7 recommends that the quantities of carbon and hydrogen in the residues be ascertained and used as the basis for calculating the calorific value.

It is necessary to determine the heat loss due to combustible matter in the various solid residues, partly for its own sake because it may be substantial and indicate a need for corrective action, but also because the proportion of unburnt fuel is a factor in determining the heat losses in the dry flue gases.

An accurate assessment of the individual losses may be physically difficult as it may for example involve opening the boiler to ensure that all residues are traced. In the past, test codes (Refs. 1 and 2) allowed test staff to adopt a simplified approach in which only a sample of the residues was taken and the total amount was calculated from the fuel burnt and the proportion of ash in the analysis of the fuel. This may still be done in approximate calculations, especially in cases where it might reasonably be assumed that all the residues are in one form, but it is more accurate to determine each loss by measurement. Where only one particular form of residue is especially difficult to measure (the quantity passing up the chimney, for example) it is permissible to treat this as the balance of the total mass, calculated as the difference between the theoretical quantity and the sum of the measured quantities. The procedure for doing this is explained in Appendix 19, and the worked example which is included shows how the production rate of unweighed residue R66 for the example in Appendix 3 is calculated. The formulae required to calculate the losses due to unburnt combustible matter in solid residues follow, with worked examples (using data from Table A3.2) for L62, L63, L66 and L67, the only losses included in the example provided in Appendix 3.

(a) Loss due to unburnt combustible content of riddlings
This loss is included only if the riddlings, which are wholly or largely raw fuel, are not refired.

$$L61 = \frac{R61 \times C61 \times Q61}{H1}$$

(b) Loss due to the unburnt combustible content of ash and clinker

$$L62 = \frac{R62 \times C62 \times Q60}{H1}$$

Using the data given in Appendix 3,

$$L62 = \frac{0.102\,34 \times 12.132 \times 33\,820}{36\,974}$$

$$= 1.1366\%$$

(c) Loss due to the unburnt combustible content of grit and dust

$$L63 = \frac{R63 \times C63 \times Q60}{H1}$$

Using the data given in Appendix 3,

$$L63 = \frac{0.029 \times 53.52 \times 33\,820}{36\,974}$$

$$= 1.4197\%$$

(d) Loss due to the unburnt combustible content of carried over fluidized bed material and ash

$$L64 = \frac{R64 \times C64 \times Q60}{H1}$$

(e) Loss due to the unburnt combustible content of mill rejects.
This loss applies only when the mill is included within the boundary of the system under test:

$$L65 = \frac{R65 \times C65 \times Q65}{H1}$$

(f) Loss due to unburnt combustible matter in unweighed solid residues

$$L66 = \frac{R66 \times C66 \times Q60}{H1}$$

The calculation of R66 is explained in Appendix 19.
Using the data given in Appendix 3,

$$L66 = \frac{0.007\,55 \times 60 \times 33\,820}{36\,974}$$

$$= 0.4143\%$$

(g) Total loss due to unburnt combustible matter in solid residues

$$L67 = L61 + L62 + L63 + L64 + L65 + L66$$

where L65 is included only if the pulverizing mill is considered to be within the unit boundary.

From the preceding calculations,

$$L67 = L62 + L63 + L66$$
$$= 1.1360 + 1.4200 + 0.4140$$
$$= 2.9706\%$$

Similar expressions to some or all of the foregoing are to be found in all of the test codes Refs. 1–7.

In passing, it is worth noting that MCR, the total mass of unburnt carbon in solid residues per kilogram of fuel, which is used in calculating L41, L42 and L43, is given by

$$MCR = \frac{(R62 \times C62) + (R63 \times C63) + (R66 \times C66)}{100 \times B11}$$

$$= \frac{(0.102\,34 \times 12.132) \times (0.029 \times 53.52) + (0.007\,55 \times 60)}{100 \times 1.33}$$

$$\approx 0.0244 \text{ kg/kg fuel}$$

4.4.13   *Losses due to sensible heat in solid residues*

The amount of sensible heat in solid residues discharged from certain types of furnace may be appreciable, as for example when the ash is discharged as a molten slag. In other cases the amount of sensible heat may be insignificantly small, unless combustion conditions are at fault. If for any reason there are significant amounts of sensible heat in any of the solid residues, they must be allowed for in the heat account. The equipment normally used for ash conveying at the test site, whether it be mechanical, pneumatic or hydraulic, may render collection and measurement of the ash difficult or even impossible. Test codes give clearance to test personnel to suspend the use of such equipment if necessary or convenient, but this should not be taken to imply that measurement of the heat in the ash itself may be disregarded.

There are three methods of assessing the losses due to sensible heat in solid residues:

(i) If, when leaving the unit in the dry condition, all of the residue is "black", that is, with no visible radiation being emitted, then the heat loss in the entire mass of residue from the fuel is estimated to be at most 0.2%. Alternatively, if only one or more of several types of residue is in this condition, the estimated percentage heat loss within it is taken as being in the same proportion to the total loss (0.2%) as its mass is to the total mass of residue.

(ii) The heat loss from the residue is calculated directly from physical data. This requires a knowledge of the mass, temperature, and specific heat capacity of the residue. This procedure is feasible only in a situation where the residue is (or can be) handled from the unit in the dry condition.

(iii) The heat loss from the residue is calculated from measurements of the change of state and/or temperature of the mass of water used to quench or convey it.

In units which discharge quantities of residue from several points, a combination of methods (i), (ii) and (iii) may be adopted.

The formulae required for the method outlined in (ii) and (iii) follow, with worked examples (using data from Appendix 3) for L72, L73 and L76, these being the only losses included in the example provided.

(a) Loss due to sensible heat in riddlings

$$L71 = \frac{R61 \times S61 \times (T61 - T1) \times 100}{H1}$$

(b) Loss due to sensible heat in ash and clinker

$$L72 = \frac{R62 \times S62 \times (T62 - T1) \times 100}{H1}$$

Using the data given in Appendix 3,

$$L72 = \frac{0.102\,34 \times 0.67 \times (300 - 20) \times 100}{36\,974}$$

$$= 0.051\,93\%$$

(c) Loss due to sensible heat in grit and dust

$$L73 = \frac{R63 \times S63 \times (T63 - T1) \times 100}{H1}$$

Using the data given in Appendix 3,

$$L73 = \frac{0.029 \times 0.67 \times (190 - 20) \times 100}{36\,974}$$

$$= 0.008\,934\%$$

(d) Loss due to sensible heat in fluidized bed material and ash

$$L74 = \frac{R64 \times S64 \times (T64 - T1) \times 100}{H1}$$

(e) Loss due to sensible heat in mill rejects

$$L75 = \frac{R65 \times S65 \times (T65 - T1) \times 100}{H1}$$

L75 is included only if the mill is included within the unit boundary.

(f) Loss due to sensible heat in the unweighed residue

$$L76 = \frac{R66 \times S66 \times (T66 - T1) \times 100}{H1}$$

Using the data given in Appendix 3,

$$L76 = \frac{0.007\,56 \times 0.67 \times (150 - 20) \times 100}{36\,974}$$

$$= 0.001\,781\%$$

Considering next the approach previously described under (iii), the following losses may replace one or more of the losses L72 to L76 inclusive.

(g) Loss due to sensible heat in ash cooling water in tanks or troughs

$$L78 = \frac{W79 \times S51 \times (T79 - T78) \times 100}{H1}$$

(h) Loss due to evaporation of ash cooling-water

At its most basic level, the evaporation of water may follow simply from the pouring of water over hot ash after firebeds have been cleaned prior to sampling and removal of the ash for laboratory determination of combustible content. In other cases, it may happen that in ordinary operation of hydraulic or water-submerged ash-handling equipment, some loss of mass of water occurs. If this is sufficiently significant to be measurable it represents heat loss:

$$L79 = \frac{(W78 - W79) \times [S51(T10 - T78) + J10] \times 100}{H1}$$

Note that if the design of the unit is such that the evaporated water enters and becomes part of the flue gases, L79 does not apply because the vapour subsequently appears in the flue gas. The loss due to W79 passing out in the flue gas is then given by L57, with W57 equivalent to (W78 − W79) (see 4.4.11). The loss due to sensible heat in solid residues is treated in detail in Refs. 4, 5 and 7; Refs. 1 and 2 advise that it is too small to be included separately in the heat account and include it in the balance of the account with the radiation loss; Ref. 6 does not refer to it but does include "unmeasured losses" with the heat emitted from the surface of the unit; Ref. 3 does not refer to it at all.

It will be noted that in the worked example the sum of the losses due to sensible heat in the solid residues is only

$$0.0519 + 0.0089 + 0.0018 = 0.0626\%$$

4.4.14   *Loss due to radiation, conduction and convection from unit*

The loss due to radiation conduction and convection from the surface of the unit (often, for brevity, referred to simply as the "radiation loss") is difficult to determine from test data. The difficulty arises in measuring temperatures over the extensive and often complex surfaces of large units. A further problem is that on certain solid fuel boilers of the "open bottom" type, heat loss may also take place downwards, through the foundations.

When it is possible to carry out a comprehensive boiler test, deriving input, output and individual losses, it is usual to treat the radiation loss as the balance of the account or lump it together with some other losses as "radiation and other unmeasured losses". This procedure is not feasible in indirect forms of test, when either the input or output is measured and obtained as the difference in the heat account.

It has become common practice in test codes when dealing with units of standard configuration, and having specified types of insulation, to assume a value for a radiation loss at maximum unit output. Alternatively, simple formulae are provided enabling this value to be calculated, and based upon the areas of water-backed and gas-backed surfaces, and the type of insulation.

Obviously, these methods are questionable unless backed by experimental data, but for units of reasonably standard configuration, confirmatory evidence for the assumed radiation losses has been obtained in some instances, more especially in cases where detailed and comprehensive testing is possible, that is, when firing with gaseous or liquid fuels. The radiation losses obtained experimentally using these fuels may reasonably be assumed to apply also in the case of solid fuel units of similar geometry.

Whichever method is used to obtain it, the value of the radiation loss at maximum boiler output represents a heat loss which is normally regarded as constant over the entire operating range of the boiler because the surface temperatures are practically constant regardless of output. The percentage loss increases in inverse proportion to boiler output, so if there is a 1% radiation loss at 100% of unit output, the loss would be 2% at 50% of output, and so on.

4.5   Thermal efficiency

4.5.1   Calculation of thermal efficiency by direct method

The thermal efficiency of the unit is derived from H1 and H35.

$$Z = \frac{H35 \times 100}{H1}$$

Expressions of this type appear in Refs. 1, 2, 4, 5, 6 and 7. Ref. 3 deals only with the indirect method. Using the data provided in Appendix 3 for the unit considered as an example,

$$Z = \frac{30\,721.031 \times 100}{36\,974}$$

$$= 83.0882\%$$

4.5.2   *Calculation of thermal efficiency by indirect methods*

(a) Output unknown

In this form of test the output is derived after calculating the thermal efficiency, as follows:

$$100 = Z + (\text{sum of all losses})$$

from which Z is obtained, then the output

$$H35 = \frac{H1 \times Z}{100}$$

Similar procedures are given in Refs. 3, 4, 5 and 7. In the unit considered as example,

$$100 = Z + L41 + L42 + L51 + L67 + L72 + L73 + L76 + L85$$

and therefore

$$Z = 100 - (6.1555 + 2.0137 + 4.7094 + 2.9706 + 0.0519$$
$$+ 0.0089 + 0.0018 + 1.000)$$
$$= 83.0882\%$$

The output is then

$$H35 = \frac{36\,974 \times 83.0882}{100}$$

$$= 30\,721.031 \text{ kW}$$

(b) Input unknown

This form of test is more or less peculiar to solid fuel units, as it is only the solid fuel firing rate which may be difficult to measure due to the comparative rarity with which precision weighing equipment is installed for this purpose.

Unfortunately, the firing rate is also an important factor in determining the losses due to combustible matter in solid residues and hence also the dry gas losses, so that this unknown quantity, $B11$, is a factor on both sides of the input/output equation. This has been ignored in Ref. 3, where no advice is provided on how to derive the loss due to combustible matter in solid residues in the case when the heat input is unknown. Ref. 5, too, gives no advice on this point, although the claim is made that the indirect method of determining the efficiency is feasible when the heat input is unknown.

It is possible, as recommended in Ref. 4, to insert measured and calculated data into the overall input/output equation and solve for the firing rate, $B11$, but the process is a laborious one. It is much simpler to program the test formulae on a computer, and then, together with all

Table 4.2   Heat account on the basis of the gross calorific value for the test data in Appendix 3

Useful output and losses from unit as percentages of heat input		
Heat flow	Symbol	%
Useful output (efficiency)	Z	83.0882
Sensible heat in dry flue gas	L41	6.1555
Unburnt carbon monoxide in flue gas	L42	2.0137
Sensible and latent heat in moisture from fuel	L51	4.7094
Unburnt combustible matter in solid residues		
Ash and clinker	L62	1.1366
Grit and dust	L63	1.4197
Unweighed residues	L66	0.4143
Sensible heat in solid residues		
Ash and clinker	L72	0.0519
Grit and dust	L73	0.0089
Unweighed residues	L76	0.0018
Radiation, conduction and convection from unit	L85	1.0000
		100.0000
Input to unit	H1	36 974.000 kW
Output from unit	H35	30 721.031 kW

known data, insert an approximate value for the firing rate B11. The computer can then carry out iterative (repeated) calculations, varying the value of B11 as necessary until the calculated value for the heat output H35 matches the measured value. Ref. 7 suggests the use of an iterative method when the fuel rate is unknown but does not detail the procedure.

4.6   Heat account

Use of formulae of the type presented in this section (or as many of them as are applicable to the unit actually under test) will provide the data which comprise the heat account. Refs. 1, 2, 3, 4, 5 and 6 indicate these data presented in tabular form, and this is a convenient way of summarizing the principal results obtained.

Table 4.2 shows a heat account derived from the data provided in Appendix 3. It should again be emphasized that in an actual heat account the percentages would not be declared beyond two decimal places.

5. Test measurements

5.1 Pre-test decisions

It is reasonably straightforward to follow the procedure given in any of the test codes to which reference has been made. With the ready availability of computers and calculators the processing of test data is also simple. The major problems associated with testing stem from the need to achieve general agreement on all aspects of the test among the various parties involved and the practical and financial difficulties involved in obtaining accurate data. To avoid costly errors, time-wasting and disputes, it is necessary that all technical and financial policy decisions should be agreed by all parties prior to the commencement of the test, relating in particular to such matters as selection of test method, establishment of the unit boundary, operation of the plant, choice of instruments, deployment of staff, taking of readings, and responsibility for costs.

Test codes provide guidance on such matters and, as might be expected, those particularly concerned with the testing of large units are most explicit. However, even the testing of small units may be considered expensive in proportion to their capital costs, and those concerned with such tests should derive benefit from a study of the detailed advice given for larger and more complicated units in, for example, Ref. 4, where the advice is provided under "Guiding Principles".

It is for those working to a particular test code to follow the advice given in it, but in this section general consideration will be given to those particular aspects that directly affect the accuracy of the measurements taken.

5.2 Pre-test examination and operation

An obvious requirement prior to a test is the need to carry out a physical examination of the unit to ensure that it is mechanically sound in the sense that ductwork is gas-tight and that pressure parts are properly sealed, with no drips of water or wisps of steam escaping to atmosphere.

The unit must then be operated at the desired test output for a period prior to the test. This is necessary to establish that burners and stokers are capable of giving clean combustion in accordance with environmental legislation, that fans, pump and fuel-handling equipment are capable of supplying the unit with the mass flows appropriate to the proposed test loads, and that test staff and test instrumentation are capable of performing satisfactorily. This period of operation, which is especially relevant to new plant, is concerned only with establishing the feasibility of carrying out the tests. A separate period of pre-test operation is in any case required to establish that steady-state conditions have been attained.

5.3 Steady state

Ideally, the unit should be working without fluctuations in output at the chosen test rating prior to and during the test period. The efficiency is likely to vary according to the output and will be declared for a given output, so for the result to have significance it should apply to steady operating conditions. Steady state is a condition which is defined in various ways in test codes (Table 5.1). It can most easily be attained in a test station, where heat exchangers or cooling towers can be installed to serve as heat sinks and thereby provide a steady load.

Such test stations are, however, usually fairly small in size, and to have an available heat load as high as 5 MW is exceptional, so on-site testing cannot be avoided in the case of a larger unit — or indeed any unit, when it is necessary to check upon its performance during its working life. This means that the only load which can be applied is that required by the heating, process or power requirement on-site. At the time of the test, this may be less than the load desired, and it may not be particularly steady. Such problems are often capable of solution simply by pre-planning. Testing in winter, or taking other units off-line, may enable the test unit to be operated at as high an output as required. A study of load patterns may indicate periods when demand is predictably steady. When steady output is possible only for short periods this may influence the choice of test, as explained under 5.4.

If, in the course of a test, some variations occur in the test conditions, these should be noted in the test report, but the declared results may, by agreement, be permitted to stand if the variations are thought not to be significant. When test conditions are more seriously erratic, the average results may still be to some degree informative, but the level of accuracy cannot be assumed.

5.4 Test duration

Provided that accurate test data are obtained, the method followed in calculating thermal performance will hold good regardless of the length of the test period, but the declared results are mean values for that period, so steady operation is necessary if the results are to be related meaningfully to a particular output. Units are designed to give optimum efficiency at or about their maximum continuous rating, although modern designs maintain high efficiency over quite a wide range of output. However, some variation in thermal efficiency must be expected with changes in load pattern, and it follows that efficiencies declared over long periods such a day, a week, a heating season, or a year will almost certainly be lower than the best possible efficiency obtained at maximum

Table 5.1 Factors involved in steady-state requirements in test codes

Test code	Fuel	Feed or return water	Steam or flow water	Flue gas	Other
Refs. 1, 2		T21	T26, W26		Drum level, draught at exit, combustion air flows
Ref. 3		T21	P26, T26, W26 ΔP26 < ± 6%	ΔT2 10°C/h, VCO2, VO2 Δ(T2 − T1) < ± 6%	ΔH35 < ± 5%
Ref. 4					
Ref. 5	B11	W21	P26, T26, W26, W29		Fuel level on grate, drum level
Ref. 6	B11	T21, W21	P26, T26, W26	ΔT2 < 5°C	ΔH35 < ± 1%, firebed conditions

load over a short test period. Thus in practice the shortest test period consistent with obtaining accurate readings is favoured for obtaining a measure of optimum output and efficiency. Other factors in favour of short tests are of course the high cost of testing and the difficulty, in the case of on-site tests, of arranging steady load conditions over an extended period.

As stated in 5.3, test codes require a unit to be operating under steady-state conditions for a period prior to the test period. This ensures that all possible variables have been stabilized. Similarly, a post-test period may be required to demonstrate that the conditions were not on the point of becoming erratic as the test ended. Both of these periods may be specified, but allowance may be made for including them in the test period, by agreement (Table 5.2).

Advice on the length of the actual test period varies considerably, according to which test code is being followed. This is regrettable, and suggests that practical research is required to establish how this factor affects test results and which of the various recommended periods can be justified. The discrepancies in the advice given apply regardless of fuel. For example, in the case of direct tests on units burning oil or gas, Ref. 6 specifies a test period of 30 min, Ref. 4, 2 h, and Refs. 5 and 7, 4 h (Table 5.2). It seems particularly inconsistent that such differences should exist between test codes, even when dealing with tests on units fired by fuels which are easily measured.

In the case of units fired by solid fuel on mechanical stokers, all test codes recommend relatively long test periods, although still with varia-

Table 5.2 Test durations recommended in test codes

Period	Firing	Minimum test period (in h) for direct (D) and indirect (I) tests											
		Refs. 1, 2		Ref. 3		Ref. 4		Ref. 5		Ref. 6		Ref. 7	
		D	I	D	I	D	I	D	I	D	I	D	I
Test	Oil/gas			1		2	2	4	4	0.5		4	4
	Waste heat											4	4
	Solid fuel												
	PF-direct firing	4				4	2	4	4			4	4
	PF-indirect firing	4*						4*					
	Cyclone							4*				4	4
	Mech. stoker (steady firing)	6		2		4	2	6	4	6		10	4
	Mech. stoker (cyclic firing)	8				8	2			4	4	24	4
	Fuels with variable cal. val.					6	2						
	Electric					1							
Pre-test	All types	1/24		1		1	1	1	1	1		1/3	1/3
Post-test	All types	1†				1†	1†	1†	1†	1			

*Longer period possibly required to obtain accurate measurement of fuel quantity.
†May, by agreement, be omitted.

tions between one and another (Table 5.2). Solid fuel can be measured to a satisfactory level of accuracy by weighing, but the extended test periods recommended for those tests in which the fuel is measured are intended to reduce the error caused by variations, over the test period, of the quantity of fuel in the system from the point of weighing up to and including the firebed. Such variations can arise from changes in the level of fuel in hoppers and furnaces, or slight changes in density during the passage of the fuel through chutes or conveyors.

Errors of this type are most likely when operating combustion equipment such as the manually de-ashed underfed stoker which functions on a non-continuous cyclic basis due to the condition of the firebed changing with the gradual build-up of ash, which is cleared periodically. Tests on units fitted with such equipment are particularly long and in contractual tests are likely to extend over the entire period (between de-ashing operations) for which the unit is capable of giving rated output.

5.5   Variations in normal working procedure

It is generally accepted that occasional operations such as sootblowing and blowdown should be suspended over the test period for the sake of simplicity. If the blowdown system is automatic and continuous, allowance for the heat flow is made in the heat account.

The operation of ash-handling equipment may also be suspended if this simplifies the collection of the residues.

5.6   Instruments

The various test codes give guidance on the accuracies considered necessary when carrying out test measurements, and suggest various types of instruments which may be used to achieve these. Such guides should never be regarded as comprehensive because allowance must be made for innovation in the field of instrumentation.

Generally, it will be found that whatever the parameter that needs to be measured, there is some scope for choice in selecting instruments, particularly if some latitude is available in the selection of the measurement tolerance. This is indeed the case, because, as reference to Appendix 21 will show, in a calculation of thermal efficiency the overall accuracy is influenced by several individual measurement accuracies, and — within reason — one measurement can be allowed to be fairly coarse if the others have closer tolerances.

This is particularly relevant when considering the more difficult measurements required in a test. Coal, for example, is difficult to sample and the overall accuracy for the determination of calorific value may only be ± 1.2%. This can be accepted simply by having close tolerances on the measurement of fuel mass, and on the determination of the heat flow in the steam or water.

5.6.1 *Simplicity in use*

As a general recommendation it is always preferable to restrict any possible error in taking a measurement to the instrument itself and reduce or remove the human element in amassing test data. Therefore, cost permitting, it is better to have a digital display than require test staff, who may be tired or preoccupied by other duties, to note needle or mercury positions relative to graduations and then transcribe these to log sheets. It is even better to have the data logged automatically by recorders. This permits accurate logging of data at short (say, 5 min) time intervals, making it much easier to establish true averages over the test period.

5.6.2 *Calibration*

Test staff can be expected to use instruments in the manner specified by test codes, related standards, and makers' instructions, but probably the most important consideration is that they should avoid the temptation to assume the accuracy of any instrument. A test is pointless if the accuracy of the measuring equipment is in doubt and the only way to avoid having results challenged on this basis is to have every instrument calibrated at least prior to and also, if required, after each test.

It is not uncommon for some instruments to be used uncalibrated — orifice plates may, for example, have an assumed accuracy attributed to them when they are manufactured, installed, and used in a prescribed manner. For this reason, various instruments may have two accuracies ascribed to them, "calibrated" and "uncalibrated", with the latter the coarser of the two. Nonetheless, calibration is always to be recommended, especially for contractual tests, because it not only ensures a better accuracy but also eliminates a possible source of contention should any test results be disputed. If calibration of instruments is beyond the ability of those involved in the test, independent test laboratories, specialist firms or the equipment suppliers should be requested to carry out the work.

Also relevant to this aspect is the working life of instrument components. It may certainly be deemed inexpedient to use equipment of exceptional fragility because a boilerhouse does not usually permit the level of care associated with laboratories. Even the sturdiest equipment does not remain in pristine condition indefinitely and examination, followed by any necessary refurbishment or replacement, is necessary from time to time and certainly prior to a test.

5.6.3 *Cost*

No-one can disregard the question of cost in relation to instrumentation but in general there is an inverse relationship between cost and scale and in the case of larger plant it is easier to justify the cost of sophisticated test equipment, which is smaller in cost relative to plant and fuel costs. For the same reason it is common to find test equipment built-in to larger plant for continuous monitoring of performance, whereas even fairly simple instruments may be sparse on small units.

Having regard to marketing pressures it would be unfair to criticize equipment suppliers or contractors for concentrating on lowest capital cost when new boiler plant is under consideration, but a purchaser cannot afford to ignore operating costs and the contribution that accurate test results can make towards reducing the amount of fuel burned.

In the same context, when new installations are being planned, provision for taking the test measurements should be examined by the prospective plant owner. For example, any fluid for which the flow or analysis would subsequently be required should be routed at some point through a straight length of piping or ductwork to accommodate the sensing elements. Relevant standards provide guidance. Even simple access to instrumentation points requires consideration and this is not always given.

5.6.4   *Typical accuracies*

In some test codes, advice is provided on how close the tolerance should be on specific measurements such as feed water flow, flue-gas temperature and so on.

This should be considered in relation to the comment under 5.6 because if the accuracies achievable with the instruments are known, the user may prefer to exercise his own choice and, depending on the fuel used for other factors, may prefer to allow extra tolerance on some measurements and compensate for this with closer tolerances on others.

However, it should be appreciated that the achievable accuracy of an instrument itself does not always mean that the measurement being sought is obtainable to the same accuracy. An obvious example is the measurement of flue-gas temperature, where the accuracy of the thermocouple is only one factor, others being the variation which may occur over the width of the duct and possible radiation to and from the duct walls.

Another consideration is that the measurement of some quantities depends upon more than one instrument. For example, a flow meter will have an accuracy specified for it, but this will be for some particular fluid temperature and therefore the ultimate accuracy of the measured quantity of fluid will depend also upon the measurements of the actual fluid temperature, which is then used in association with a correction curve or formula to obtain the actual flow.

Table 5.3 lists typical accuracies which may be expected from the instruments which are currently available but the following points should also be noted.

(a) Time

The time factor in tests can be discounted as a source of inaccuracy. The period of any test is sufficiently long and the accuracy of the common quartz timekeeper so high that any timed period can be taken as exact.

Table 5.3 Typical accuracies of measurement

Parameter	Category	Measuring device	Typical accuracy ±%
Time		Quartz clock or watch	0
Mass		Platform scales	0.1
		Batch weigher	0.2
		Belt weigher	0.5
		Load cells under bunkers	0.2
		Weighbridge (commercial)	0.5
Fluid flow	Water	Orifice plate or nozzle	0.5 (2*)
		Venturi	1
		Variable area	2
		Electromagnetic	1
		Vortex shedding	0.5
		Turbine	0.25
	Fuel oil	Rotary piston	1
		Sliding vane	0.5
		Venturi	1
	Steam	Orifice plate or nozzle	0.5 (2*)
		Venturi	1
	Gaseous fuel	Orifice plate or nozzle	0.5 (2*)
		Venturi	1
		Pitot-static	2
		Positive displacement	0.5
Flue-gas analysis	VCO_2	Orsat absorption type	0.1
		Thermal conductivity	0.2
		Compact absorption type	0.3
		Infra-red	0.2
	VCO	Colorimetric	10
		Orsat absorption type	10
		Infra-red	0.2
		Electrochemical	5
	VO_2	Orsat absorption type	0.2
		Paramagnetic	0.1
		Electrochemical	0.2
		Fuel cell	0.2
Temperature		Mercury in glass thermometer	0.5°C
		Mercury in steel thermometer	1°C
		Thermocouple	1°C
		Resistance thermometer	0.1°C
Calorific value	Oil/gas	Calorimeter (+ sampling)	0.6
	Solid fuel	Calorimeter (+ sampling)	1.2
Pressure	Low (air/gas)	Diaphragm	0.5
	High (steam, etc.)	Bourdon	1

*Uncalibrated

(b) Mass

Weighing is the only accurate method of determining the mass of solid fuel and is also used, especially in test houses, for measuring oil fuel and sometimes feed or circulating water.

When testing smaller units it is reasonably convenient to measure quantities of coal by manually filling and wheeling trolleys of coal on to platform scales, before emptying them into the chutes or conveying systems feeding fuel to the units.

Unfortunately, the cost of weighing equipment is disproportionately high in relation to the capital costs of the many medium to large units burning upwards of say 1t/h. For this reason, weighing equipment is often not installed so the testing or setting-up of such units is rendered much more difficult. Indirect testing is a possible alternative, though not if the unit is producing saturated steam because this cannot be accurately sampled.

It may be decided to fit load cells under service bunkers, which is usually acceptable, or under coal storage silos, which is more questionable because a test may involve a fairly small difference in a large overall total weight of fuel. Furthermore, care would be necessary to ensure that the loss of weight did not include moisture drained from the stored coal. In an average "over the day" (or week, or month) test it may be agreed to accept delivery vehicle weighbridge readings provided that a wider accuracy in the determination of unit output and efficiency is acceptable.

It is not uncommon to find volumetric measurement of coal resorted to in difficult situations, with bunker levels, discharge from screw conveyors, and so on, being monitored. Even with calibration, this indirect estimation of mass cannot be accepted as reliable and, again, any test result declared on this basis must be recognized as being to a lower level of accuracy than if it were based on direct measurement.

Indirect measurement of the mass of oil fuels by use of calibrated tanks is acceptable, provided that any corrections for temperature variations are rigorously applied.

(c) Flow

The flow measurement devices in Table 5.3 are those usually used in test work, but any other instrument of the required accuracy and suitable for the working conditions may be used. Oil and water flow may also be measured by weighing over a timed period or by use of tanks, calibrated volumetrically.

Where possible, it is preferred to measure the feed water into a unit rather than the steam out of it, because the meter is then working with a cooler and denser fluid. If the feed water has a pulsating flow, as when reciprocating pumps are used, meters of the differential pressure type (orifice plates, nozzles and venturis) must not be used.

(d) Flue-gas analysis

The formerly ubiquitous Orsat, and similar absorptive analysers, have largely given way to other techniques for reasons of simplicity, robustness and speed in use, although compact absorption CO_2 and O_2 analysers are still popular for simple, quick monitoring of performance.

Except in the case of infra-red gas analysers, which direct a beam of infra-red radiation across the flue-gas duct, sampling is required prior to analysis and this may affect the overall accuracy of the determination. Any stratification of the flue gas is allowed for by traversing the duct with both pitot and sampling probes, "weighting" the individual analysis readings to allow for the velocity measurements, then establishing an acceptable mean point.

(e) Temperature

Of the instruments commonly used, mercury-in-glass and mercury-in-steel types are likely to be employed now only on small units.

The high accuracy of the resistance thermometer is of particular benefit when hot-water units are tested and especially if the temperature difference between the flow and return is fairly low, as it might be in the case of low- or medium-temperature heating systems. High accuracy in temperature measurements is then important if the overall determination of efficiency is to be within acceptable limits.

For accurate measurement of the temperature of flue gas the duct has to be traversed in similar fashion to that required for analysis, in order to establish a mean temperature point. If one point is found to be at that mean, the thermocouple probe may be left at that point provided the working conditions of the unit remain stable. Fine wire thermocouples are favoured for rapid results and to minimize radiation between the probe and the duct walls. Ref. 3 advises that such radiation is also reduced by lagging the ductwork up to a point at least one duct diameter downstream of the probe.

(f) Calorific value

In the case of solid fuel, the overall accuracy of the determination of calorific value is more seriously affected by sampling errors than is the case with gaseous or liquid fuels and this is reflected in the figure given in Table 5.3. The sampling of solid fuel must conform to an established, standard procedure such as that given in Ref. 13.

Appendix 1. Symbols and units

Table A1.1 lists the symbols and units used in the calculations. For simplicity, and as an aid to anyone wishing to use them in computer programs, none of the symbols representing physical quantities include Greek letters, lower-case letters, subscripts, superscripts or indices. Any numerals or letters used in parenthesis after symbols to distinguish different categories, for example, MH2(1), are explained at the point of use.

The equivalent imperial units have been listed as these are still widely used. It should however be noted that formulae employing these may differ from those derived for metric units in respect of the constants used.

Table A1.1   Symbols and units

Symbol	Quantity	Metric	Imp
		Units	
Discharge of pure ash in solid residues			
A61	Pure ash in riddlings	kg/s	lb/h
A62	Pure ash in ash and clinker	kg/s	lb/h
A63	Pure ash in grit and dust	kg/s	lb/h
A64	Pure ash in fluidized bed material and ash	kg/s	lb/h
A65	Pure ash in mill rejects	kg/s	lb/h
A66	Pure ash in unweighed residue	kg/s	lb/h
Rate of firing			
B11	Fuel by mass (or mass flow of waste gas in waste heat boiler)	kg/s	lb/h
B11V	Gaseous fuel, by volume, at temperature T5 and pressure P5	m^3/s	ft^3/h
Combustible contents of solid residues (dry basis)			
C61	Riddlings	%	
C62	Ash and clinker	%	
C63	Grit and dust	%	
C64	Fluidized bed material and ash	%	
C65	Mill rejects	%	
C66	Unweighed residue	%	
Dryness fraction of steam			
D26	Dryness fraction of steam at discharge from unit	kg/kg	lb/lb
Specific enthalpies			
E1	Air at temperature T1 and pressure P1	kJ/kg	Btu/lb
E10	Water at temperature T1	kJ/kg	Btu/lb

Table A1.1 — *Continued*

Symbol	Quantity	Units	
		Metric	Imp
E13	Steam or water injected into furnace at temperature T13 and pressure P13	kJ/kg	Btu/lb
E14	Air from separate source at temperature T14 and pressure P14	kJ/kg	Btu/lb
E17	Waste gas at inlet to unit	kJ/kg	Btu/lb
E18	Waste gas at discharge from unit	kJ/kg	Btu/lb
E19	Waste gas at temperature T1 and pressure P1	kJ/kg	Btu/lb
E21	Feed water at temperature T21	kJ/kg	Btu/lb
E22	Main steam attemperator spray water at temperature T22	kJ/kg	Btu/lb
E23	Reheated steam attemperator spray water at temperature T23	kJ/kg	Btu/lb
E25	Blowdown (drum water) at temperature T25	kJ/kg	Btu/lb
E26	Steam or water at discharge from unit at temperature T26 and pressure P26	kJ/kg	Btu/lb
E27	Steam returned to unit for use on ancillaries	kJ/kg	Btu/lb
E28	Steam returned to unit for reheating at temperature T28 and pressure P28	kJ/kg	Btu/lb
E29	Reheated steam discharged from unit at temperature T29 and pressure P29	kJ/kg	Btu/lb
E31	Water at entry to condensing recuperator	kJ/kg	Btu/lb
E32	Water at outlet from condensing recuperator	kJ/kg	Btu/lb
Gaseous fuel analysis (volumetric basis)			
FCO	Carbon monoxide	%	
FCO2	Carbon dioxide	%	
FCH4	Methane	%	
FC2H2	Acetylene	%	
FC2H4	Ethylene	%	
FC2H6	Ethane	%	
FC3H8	Propane	%	
FC4H10	Butane	%	
FC5H12	Pentane	%	
FC6H14	Hexane	%	
FC7H16	Heptane	%	
FC8H18	Octane	%	
FH2	Hydrogen	%	
FH20	Water	%	
FN2	Nitrogen	%	
FO2	Oxygen	%	
Heat inputs to unit			
H1	Total	kW	Btu/h
H11	Total calorific value of the fuel fired	kW	Btu/h
H12	Sensible heat in fuel	kW	Btu/h
H13	Steam or water injected into furnace	kW	Btu/h
H14	Air supplied from a separate source	kW	Btu/h
H15	Substances injected into the furnace	kW	Btu/h
H16	Mechanical energy of ancillaries	kW	Btu/h
H17	Heat from waste gas	kW	Btu/h
H18	Electricity	kW	Btu/h
Heat flows in heat carrier			
H21	Feed water	kW	Btu/h
H22	Main steam attemperator spray water	kW	Btu/h
H23	Reheated steam attemperator spray water	kW	Btu/h

Table A1.1 — *Continued*

Symbol	Quantity	Units Metric	Units Imp
H24	Variation in water level	kW	Btu/h
H25	Blowdown	kW	Btu/h
H26	Main steam or water discharge	kW	Btu/h
H27	Steam returned to unit for use on ancillaries	kW	Btu/h
H28	Steam returned to unit for reheating	kW	Btu/h
H29	Reheated steam discharged from unit	kW	Btu/h
H31	Water at entry to condensing recuperator	kW	Btu/h
H32	Water at outlet from condensing recuperator	kW	Btu/h
H35	Total output from unit	kW	Btu/h
	Latent heat, dry saturated condition		
J10	Steam at pressure P1	kJ/kg	Btu/lb
J26	Steam at pressure P26		
J51	Steam at pressure P51	kJ/kg	Btu/lb
	Miscellaneous factors		
K1	Relative humidity of combustion air	%	
K2	Universal gas constant	kJ/(kg mol K)	ft lbf/(lb mol R)
K41	Constant in abbreviated formula for L41		
K42	Constant in abbreviated formula for L42		
K54	Water content of combustion air by mass	kg/kg	lb/lb
K64	Proportion of ash in fluidized bed material and ash		%
	Losses		
L25	Blowdown	%	
L41	Sensible heat in dry flue gas	%	
L42	Unburnt carbon monoxide in flue gas	%	
L43	Unburnt hydrocarbons in flue gas	%	
L44	Heat in waste gas at discharge from unit	%	
L51	Sensible and latent heat in moisture from fuel	%	
L52	Sensible heat in moisture from fuel	%	
L53	Steam or water injected into furnace	%	
L54	Moisture in the combustion air	%	
L55	Condensed moisture from the flue gas	%	
L56	Residual moisture in the flue gas	%	
L57	Evaporated ash cooling-water in the flue gas	%	
L61	Unburnt combustible matter in riddlings	%	
L62	Unburnt combustible matter in ash and clinker	%	
L63	Unburnt combustible matter in grit and dust	%	
L64	Unburnt combustible matter in fluidized bed material and ash		%
L65	Unburnt combustible matter in mill rejects	%	
L66	Unburnt combustible matter in unweighed residues		%
L67	Total loss due to unburnt combustible matter in solid residues		%
L71	Sensible heat in riddlings	%	
L72	Sensible heat in ash and clinker	%	
L73	Sensible heat in grit and dust	%	
L74	Sensible heat in fluidized bed material and ash	%	
L75	Sensible heat in mill rejects	%	
L76	Sensible heat in unweighed residue	%	
L78	Heat in ash cooling-water	%	
L85	Radiation, conduction and convection from unit		%

Table A1.1 — *Continued*

Symbol	Quantity	Metric	Imp
		Units	

Symbol	Quantity	Metric	Imp
L86	Radiation, convection from boilerhouse	%	
	Fuel analysis and related quantities (mass basis)		
MA	Ash		
MC	Carbon	kg/kg	lb/lb
MH2	Hydrogen	kg/kg	lb/lb
MH20	Water	kg/kg	lb/lb
MN2	Nitrogen	kg/kg	lb/lb
MO2	Oxygen	kg/kg	lb/lb
MS	Combustible sulphur	kg/kg	lb/lb
MCR	Carbon lost in solid residues	kg/kg	lb/lb
MG	Dry gas in flue gas per unit mass of fuel	kg mol/kg	lb mol/lb
MW	Water vapour in flue gas per unit mass of fuel	kg mol/kg	lb mol/lb
	Pressures		
	The pressures are gauge unless otherwise stated		
P1	Ambient air at inlet to unit (absolute)	mbar	lbf/in^2
P2	Flue gas at discharge from unit (absolute)	mbar	lbf/in^2
P5	Reference for gaseous fuel (absolute)	mbar	lbf/in^2
P13	Steam or water injected into furnace	bar	lbf/in^2
P14	Air supplied from a separate source	bar	lbf/in^2
P17	Waste gas at inlet to unit	bar	lbf/in^2
P18	Waste gas at discharge from unit	bar	lbf/in^2
P21	Feed water	bar	lbf/in^2
P22	Main steam attemperator spray water	bar	lbf/in^2
P23	Reheated steam attemperator spray water	bar	lbf/in^2
P25	Boiler blowdown (shell or drum)	bar	lbf/in^2
P26	Steam or water at discharge from unit	bar	lbf/in^2
P27	Ancillary steam for use within unit	bar	lbf/in^2
P28	Steam returned to unit for reheating	bar	lbf/in^2
P29	Reheated steam discharged from unit	bar	lbf/in^2
P31	Water supplied to condensing recuperator	bar	lbf/in^2
P32	Water discharged from condensing recuperator	bar	lbf/in^2
P51	Partial pressure of water vapour in the flue gas	bar	lbf/in^2
	Calorific values		
Q11GV	Gross calorific value of the fuel, by mass, measured at constant volume (and corrected to reference temperature T5)	kJ/kg	Btu/lb
Q11	Gross calorific value of fuel, by mass, at constant pressure and at temperature T5	kJ/kg	Btu/lb
Q11N	Net calorific value of fuel, by mass, at constant pressure and at temperature T5	kJ/kg	Btu/lb
Q11V	Gross calorific value of gaseous fuel, by volume measured at constant pressure (and corrected to temperature T5)	kJ/kg	Btu/lb
Q11VN	Net calorific value of gaseous fuel, by volume, measured at constant pressure (and corrected to temperature T5)	kJ/kg	Btu/lb
Q12	Calorific value of carbon burning to carbon dioxide	kJ/kg	Btu/lb
Q13	Calorific value of carbon burning to carbon monoxide	kJ/kg	Btu/lb

Table A1.1 — *Continued*

Symbol	Quantity	Units	
		Metric	Imp
Q15	Net exothermic/endothermic heat resulting from substances injected into furnace	kJ/kg	Btu/lb
Q42	Calorific value of carbon monoxide in flue gas	kJ/kg	Btu/lb
Q43	Calorific value of unburnt hydrocarbon gases in flue gas	kJ/kg	Btu/lb
Q60	Calorific value of unburnt combustible matter in the solid residues of combustion	kJ/kg	Btu/lb
Q61	Calorific value of unburnt combustible matter in riddlings	kJ/kg	Btu/lb
Q65	Calorific value of unburnt combustible matter in mill rejects	kJ/kg	Btu/lb
Discharge of solid residues (dry basis)			
R61	Riddlings	kg/s	lb/h
R62	Ash and clinker	kg/s	lb/h
R63	Grit and dust	kg/s	lb/h
R64	Fluidized bed material and ash	kg/s	lb/h
R65	Mill rejects	kg/s	lb/h
R66	Unweighed residue	kg/s	lb/h
Mean specific heat capacities			
S12	Unit mass of fuel	kJ/(kg K)	Btu/(lb R)
S41	Dry flue gas	kJ/(kg mol K)	Btu/(lb mol)
S51	Water	kJ/(kg K)	Btu/(lb R)
S52	Superheated steam in flue gas	kJ/(kg K)	Btu/(lb R)
S61	Riddlings	kJ/(kg K)	Btu/(lb R)
S62	Ash and clinker	kJ/(kg K)	Btu/(lb R)
S63	Grit and dust	kJ/(kg K)	Btu/(lb R)
S64	Fluidized bed material and ash	kJ/(kg K)	Btu/(lb R)
S65	Mill rejects	kJ/(kg K)	Btu/(lb R)
S66	Unweighed residue	kJ/(kg K)	Btu/(lb R)
Temperatures			
T0	Absolute temperature at freezing point of water	°C	°F
T1	Combustion air at entry to air intakes of unit	°C	°F
T2	Flue gas leaving unit	°C	°F
T3	"Wet" bulb thermometer of psychrometer at air inlet to unit	°C	°F
T4	Air at entry to boilerhouse	°C	°F
T5	Reference temperature of the calorific value (and of the volume rating in the case of a gaseous fuel)	°C	°F
T6	Ambient temperature of air in boilerhouse	°C	°F
T10	Saturated steam at atmospheric pressure	°C	°F
T11	Fuel entering unit	°C	°F
T12	Fuel entering furnace	°C	°F
T13	Steam or water injected into furnace	°C	°F
T14	Air supplied from a separate source	°C	°F
T17	Waste gas at inlet to unit	°C	°F
T18	Waste gas at discharge from unit	°C	°F
T21	Feed water	°C	°F
T22	Main steam attemperator spray water	°C	°F
T23	Reheated steam attemperator spray water	°C	°F
T25	Blowdown water	°C	°F
T26	Steam or water at discharge from unit	°C	°F
T27	Steam returned to unit for ancillary use	°C	°F

Table A1.1 — *Continued*

Symbol	Quantity	Units Metric	Units Imp
T28	Steam returned to unit for reheating	°C	°F
T29	Reheated steam discharged from unit	°C	°F
T51	Temperature (saturation) of steam in flue gas at partial pressure P51	°C	°F
T55	Condensed flue-gas moisture discharged from condensing recuperator	°C	°F
T61	Riddlings	°C	°F
T62	Ash and clinker	°C	°F
T63	Grit and dust	°C	°F
T64	Fluidized bed material and ash	°C	°F
T65	Mill rejects	°C	°F
T66	Unweighed residue	°C	°F
T78	Water entering tank or trough for cooling solid residues	°C	°F
T79	Water leaving tank or trough for cooling solid residues	°C	°F

Flue-gas analysis (by volume, dry basis)

Symbol	Quantity	Units Metric	Units Imp
VCO	Carbon monoxide	%	
VCO2	Carbon dioxide	%	
VCXHY	Hydrocarbon gases	%	
VN2	Nitrogen	%	
VO2	Oxygen	%	
VSO2	Sulphur dioxide	%	

Volumetric quantities

Symbol	Quantity	Units Metric	Units Imp
VFG	Combined dry gas and water vapour in flue gas per unit mass of fuel	m^3/kg	ft^3/lb

Mass flow rates

Symbol	Quantity	Units Metric	Units Imp
W1	Combustion air entering unit	kg/s	lb/h
W11	Solid fuel prior to on-site milling	kg/s	lb/h
W13	Steam or water injected into furnace	kg/s	lb/h
W14	Combustion air from separate source	kg/s	lb/h
W15	Substance injected into furnace	kg/s	lb/h
W21	Feed water	kg/s	lb/h
W22	Main stream attemperator spray water	kg/s	lb/h
W23	Reheated steam attemperator spray water	kg/s	lb/h
W24	Fall in drum water-level during test (negative if level rises over test)	kg/s	lb/h
W25	Blowdown	kg/s	lb/h
W26	Steam or water at discharge from unit .	kg/s	lb/h
W27	Steam returned to unit for use on ancillaries	kg/s	lb/h
W28	Steam returned to unit for reheating	kg/s	lb/h
W29	Reheated steam discharged from unit	kg/s	lb/h
W31	Water circulated through condensing recuperator	kg/s	lb/h
W55	Condensed flue-gas moisture discharged from condensing recuperator	kg/s	lb/h
W56	Residual moisture in flue gas	kg/s	lb/h
W57	Evaporated ash cooling-water in the flue gas	kg/s	lb/h
W78	Cooling-water entering solid residues tank or trough	kg/s	lb/h
W79	Cooling-water leaving solid residues tank or trough	kg/s	lb/h

Table A1.1 — *Continued*

Symbol	Quantity	Units	
		Metric	Imp
Numbers of atoms in hydrocarbon gases			
X	Number of carbon atoms in one molecule of unburnt hydrocarbon gas		
Y	Number of hydrogen atoms in one molecule of unburnt hydrocarbon gas		
Thermal efficiency			
Z	Thermal efficiency	%	

Appendix 2. Combustion reactions of fuels and related substances

Table A2.1 lists fuels and related substances, together with combustion reactions. As explained in Appendix 1, chemical symbols do not include subscript numbers.

Table A2.1 Combustion reactions — fuels and related substances

Substance	Molecular formula	Relative molecular mass	Combustion reactions (molar quantities)
Air		28.964	
Carbon	C	12.011	1C + 0.5 O2 → 1CO
			1C + 1 O2 → 1CO2
Carbon monoxide	CO	28.011	1CO + 0.5 O2 → 1CO2
Carbon dioxide	CO2	44.010	
Hydrogen	H2	2.016	1H2 + 0.5 O2 → 1H2O
Nitrogen	N2	28.013	
Oxygen	O2	31.998	
Sulphur	S	32.066	1S + 1 O2 → 1SO2
Methane	CH4	16.043	1CH4 + 2 O2 → 1CO2 + 2H2O
Acetylene	C2H2	26.038	1C2H2 + 2.5 O2 → 2CO2 + 1H2O
Ethylene	C2H4	28.054	1C2H4 + 3 O2 → 2CO2 + 2H2O
Ethane	C2H6	30.070	1C2H6 + 3.5 O2 → 2CO2 + 3H2O
Propane	C3H8	44.097	1C3H8 + 5 O2 → 3CO2 + 4H2O
Butane	C4H10	58.124	1C4H10 + 6.5 O2 → 4CO2 + 5H2O
Pentane	C5H12	72.151	1C5H12 + 8 O2 → 5CO2 + 6H2O
Hexane	C6H14	86.179	1C6H14 + 9.5 O2 → 6CO2 + 7H2O
Heptane	C7H16	100.206	1C7H16 + 11 O2 → 7CO2 + 8H2O
Octane	C8H18	114.233	1C8H18 + 12.5 O2 → 8CO2 + 9H2O
Hydrogen sulphide	H2S	34.082	1H2S + 1.5 O2 → 1H2O + 1SO2
Sulphur dioxide	SO2	64.070	
Water	H2O	18.015	

Data on relative molecular masses may vary slightly according to source. In many test codes they are rounded to whole numbers.

Many other hydrocarbon compounds may be encountered but those listed are sufficient for illustrating the method of treatment.

In carrying out calculations on combustion processes and the gaseous products of combustion, the following further data have been used:

Standard atmospheric pressure = 101.325 kPa
$$= 1.013\,25 \text{ bar}$$
$$= 14.695\,9 \text{ lbf/in}^2$$

$0°C = 273.15 \text{ K} = 492 \text{ R}$

1 kilogram mole (1 kg mol) of any gas at standard atmospheric pressure and 0°C occupies 22.414 m^3. 1 pound mole (1 lb mol) occupies 359 ft^3.

Universal gas constant = 8.314 kJ/(kg mol K) = 1545 ft lbf/(lb mol R)

Density of air at standard atmospheric pressure and 0°C = 1.293 kg/m^3
$$= 0.08072 \text{ lb/ft}^3$$

Composition of air:
    Gravimetric: 23.2% oxygen, 76.8% nitrogen
    Volumetric: 21% oxygen, 79% nitrogen

Appendix 3.   Data used in worked examples

This Appendix lists the data used in the worked examples which appear in the text and in the other Appendices. The typical fuel data in Table A3.1 are based principally upon the comprehensive lists published in Ref. 11.

The boiler test data in Table A3.2 are hypothetical, and have been selected simply to provide numerical values for the worked examples. As stated in 4.1, the data and calculated values may be shown with an unrealistic number of significant figures in order to reduce the cumulative errors that result from repeated rounding off in the stages of lengthy calculations.

Fig. A3.1 shows the boundary points of the hypothetical unit.

Table A3.3 has been included for interest. It lists the values recommended in various test codes for a number of factors and, not unexpectedly, shows that for the most part these are closely similar.

Table A3.1 Fuel data used in worked examples

Fuel state:	Solid			Liquid				Gaseous
	Coal Rank Code Number			Fuel oil class (BS 2869)				
Fuel type:	Washed duff 101	Washed smalls 601	Washed smalls 802	D	E	F	G	North Sea nat. gas
Mass analysis kg/kg fuel								
MC	0.782	0.678	0.613	0.861	0.856	0.856	0.854	
MH2	0.024	0.042	0.040	0.132	0.117	0.115	0.114	
MH2O	0.120	0.110	0.160					
MN2	0.009	0.014	0.013					
MS	0.010	0.017	0.017	0.007	0.025	0.026	0.028	
MA	0.080	0.080	0.080		0.002	0.003	0.004	
MO	0.015	0.059	0.077					
Vol. analysis m³/m³ fuel								
FCO2								0.002
FN2								0.015
FCH4								0.944
FC2H6								0.030
FC3H8								0.005
FC4H10								0.002
FC5H12								0.001
FC6H14								0.001
Gross calorific value kJ/kg fuel	29 650	27 800	25 250	45 600	43 500	43 100	42 900	
kJ/m³ fuel (at. press., 15°C)								38 620
Density kg/m³ (at. press., and 15°C)								0.723

Source: *Technical Data on Fuel* (Ref. 11)

Table A3.2 Boiler test data used in worked examples

Quantity symbol	Numerical value	Units	Quantity symbol	Numerical value	Units
A62	0.0899	kg/s	P1	1013.25	mbar (abs)
A63	0.0135	kg/s	P2	1013.25	mbar (abs)
A66	0.0030	kg/s	Q11	27 800	kJ/kg
B11	1.33	kg/s	Q42	10 100	kJ/kg
C62	12.132	%	Q60	33 820	kJ/kg
C63	53.520	%	R62	0.102 34	kg/s
C66	60.000	%	R63	0.029 0	kg/s
H1	36 974	kW	R66	0.007 55	kg/s
H35	30 721.031	kW	S41	30.6	kJ/(kg mol K)
L41	6.1555	%	S51	4.2	kJ/(kg K)
L42	2.0137	%	S52	1.88	kJ/(kg K)
L51	4.7094	%	S62	0.67	kJ/(kg K)
L62	1.1366	%	S63	0.67	kJ/(kg K)
L63	1.4197	%	S66	0.67	kJ/(kg K)
L66	0.4143	%	T1	20	°C
L67	2.9706	%	T2	150	°C
L72	0.0519	%	T51	39	°C
L73	0.0089	%	T62	300	°C
L76	0.0018	%	T63	190	°C
L85	1.0000	%	T66	150	°C
MC	0.678	kg/kg	VCO2	12.19	%
MH2	0.042	kg/kg	VCO	0.46	%
			VCXHY	0.00	%
MH2O	0.110	kg/kg	VN2	80.41	%
MN2	0.014	kg/kg	VO2	6.82	%
MS	0.017	kg/kg	VSO2	0.12	%
MO2	0.059	kg/kg	Z	83.0882	%
MA	0.080	kg/kg			
MG	0.4299	kg mol/kg			
MW	0.0269	kg mol/kg			
MCR	0.0244	kg/kg			

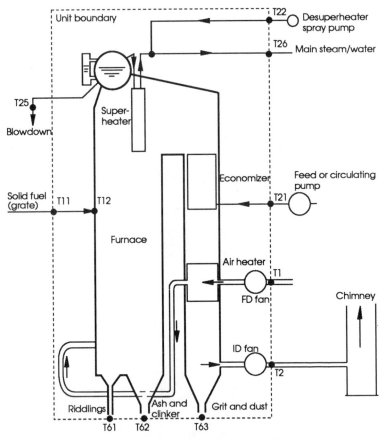

Fig. A3.1　Boundary points of unit in example

The figures actually given in the references are underlined, the equivalent figures in other units being given to aid comparison. Where there is no underlined figure in any group, this indicates that the factor has not been used directly in the reference, but a value has been derived from other factors used. See also Appendix 11.

Table A3.3 Physical data used in test codes

Quantity	Symbol	Units	Numerical values given in test codes						
			Ref. 1	Ref. 2	Ref. 3	Ref. 4	Ref. 5	Ref. 6	Ref. 7
Specific heat of dry flue gas	S41	kJ/kg mol K / kJ/kg K / Btu/lb R	30.1 / 1.006 / 0.237	30.1 / 1.006 / 0.237	30.6 / 1.011 / 0.241	30.6 / 1.011 / 0.241	30.6 / 1.011 / 0.241	30.6 / 1.011 / 0.241	30.6 / 1.011 / 0.241
Specific heat of water	S51	kJ/kg K / Btu/lb R	4.2 / 1.0	4.2 / 1.0	4.2 / 1.0	4.2 / 1.0	4.2 / 1.0	4.2 / 1.0	4.2 / 1.0
Specific heat of super-heated water vapour in flue gas	S52	kJ/kg K / Btu/lb R	2.1 / 0.5	2.1 / 0.5	2.1 / 0.5	1.88 / 0.45	1.88 / 0.45	2.1 / 0.5	
Saturation temperature of water vapour in flue gas	T51	°C / °F	100 / 212	100 / 212	100 / 212	39 / 102	25 / 77	100 / 212	
Latent heat of evaporation of water vapour in flue gas	J51	kJ/kg / Btu/lb	2256 / 970	2250 / 967	2488 / 1070	2409 / 1036	2442 / 1050	2250 / 967	
Calorific value of unburnt material (carbon) in solid residues	Q60	kJ/kg / Btu/lb	33 820 / 14 540	33 820 / 14 540	33 820 / 14 540	33 820 / 14 540	33 820 / 14 540	33 820 / 14 540	33 730 / 14 500
Specific heat of solid residues	S62-66	kJ/kg K / Btu/lb R					0.67 / 0.16		1.05 / 0.25

Appendix 4.   Conversion of solid fuel analyses

In dealing with coal or other solid fuel, it is often necessary to convert from one analytical base to another. All of the conversion factors required for this purpose will be found in Ref. 12 (part 16). Alternatively, they may be derived as the need arises, as follows.

Usually the position is that an analysis to a first base (1) is known in full, whilst the required analysis to a second base (2) is known only in respect of one or more constituents that have varied, such as moisture and/or ash.

Given this information about moisture (and/or ash) it follows that the sum of the unvarying constituents to the second base comprises

$$1 - [MH2O(2) + MA(2)]$$

Previously, the same constituents totalled

$$1 - [MH2O(1) + MA(1)]$$

Therefore, the revised proportion of any constituents originally present in (1) and still present in (2) is given by

$$\text{Proportion}\,(2) = \text{Proportion}\,(1) \times \frac{1 - [MH2O(2) + MA(2)]}{1 - [MH2O(1) + MA(1)]}$$

The general case is that any revised proportion (2) is given by

$$\text{Proportion}\,(2) = \text{Proportion}\,(1) \times$$

$$\left[1 - \frac{(\text{sum of constituent proportions that vary in second analysis})}{(\text{sum of same variable constituent proportions in first analysis})} \right]$$

Note that if percentages are used instead of proportions, 1 is replaced by 100.

As an example illustrating the method, consider the Rank 601 coal in Table A3.1, and assume that this is being used as the feedstock for manufacturing a coal–water mixture in which the water will finally amount in total to a proportion of 0.3 of unit mass. Only the moisture is a variable in this case and the conversion factor for the carbon, hydrogen, sulphur, ash and oxygen is therefore

$$\frac{1-0.3}{1-0.11}$$

$$=\frac{0.7}{0.89}$$

$$=0.7865$$

Thus, the proportions by mass of the final coal–water mixture will be

$$MC = 0.533$$
$$MH2 = 0.033$$
$$MH2O = 0.300$$
$$MN2 = 0.011$$
$$MO2 = 0.046$$
$$MS = 0.013$$
$$MA = 0.063$$

The calorific value, 27 800 kJ/kg, is converted by using the same factor, and becomes 21 865 kJ/kg.

Appendix 5.   Conversion of calorific values

(a) Solid and liquid fuels

The calorific values of these fuels are first determined on the gross constant-volume basis using a bomb calorimeter. The reference temperature is 25°C. Refs. 10 and 11 provide the following conversion factors.

For the gross calorific value on the constant-pressure basis
add 6.2 MH2 − 0.8 MO2 (kJ/kg)
or 2.6 MH2 − 0.3 MO2 (Btu/lb)
For the net calorific value on the constant-pressure basis
subtract 212 MH2 + 24.4 MH2O + 0.8 MO2 (kJ/kg)
or 91.2 MH2 + 10.5 MH2O + 0.3 MO2 (Btu/lb)

(b) Gaseous fuels

The calorific values of these fuels are normally first determined on the gross content-pressure basis using a continuous-flow calorimeter. The reference temperature is 15°C. Refs. 10 and 11 provide the following conversion factors.

For the net calorific value on the constant-pressure basis

$$\text{subtract } 18.78 \left[\text{FH2} + \frac{(\text{FCXHY} \times \text{Y})}{2} \right] \text{kJ/m}^3$$

$$\text{or } 0.49 \left[\text{FH2} + \frac{(\text{FCXHY} \times \text{Y})}{2} \right] \text{Btu/ft}^3$$

Appendix 6.   Conversion of a gaseous fuel specification from a volumetric to a mass basis

In test calculations concerned only with a gaseous fuel, the formulae can be structured to suit a fuel specified on a volumetric basis. In more general test codes it may be convenient to convert a gaseous fuel specification on a volumetric basis to one on a mass basis so that the same test formulae can apply to solid, liquid and gaseous fuels. Table A6.1 illustrates the procedure for carrying out such a conversion, the gas chosen for the example being the North Sea natural gas in Table A3.1.

The sum of column (e) is the equivalent relative molecular mass of the mixture. From the normal molar relationship therefore, 17.108 kg occupies 22.414 m^3 at standard atmospheric pressure (1013.25 mbar) and 0°C (273.15 K).

Gaseous fuel characteristics are declared at standard atmospheric pressure and 15°C (288.15 K) and at this temperature 22.414 m^3 would expand to

$$22.414 \times \frac{288.15}{273.15}$$

$$= 23.643 \text{ m}^3$$

The density of the gas at atmospheric pressure and 15°C would therefore be

$$\frac{17.108}{23.643}$$

$$= 0.7236 \text{ kg/m}^3$$

The density and other characteristics of this gas are fully specified in Ref. 11 but the above procedure may be found useful for less familiar and unlisted mixtures of gases.

The calorific value is declared as 38 620 kJ/m^3 at 15°C, which, divided by the declared density at 15°C, 0.723, gives a calorific value on a mass basis of 53 420 kJ/kg.

Table A6.1 Conversion of a gaseous fuel specification from a volumetric to a mass basis

Fuel constituent Symbol	Proportion m³/m³ fuel (a)	Compressibility factor (b)	Proportion, allowing for compressibility m³/m³ fuel (c) = (a)/(b)	Relative molecular mass (d)	Proportion of each constituent by mass kg/kg mol fuel (e) = (c)×(d)	Proportion of each constituent by mass kg/kg fuel (f) = (e)/Σ(e)	Elements in fuel kg mol/kg mol fuel			
							Carbon (g)	Hydrogen (h)	Nitrogen (i)	Oxygen (j)
FCO2	0.002	0.9943	0.0020	44.010	0.0885	0.0052	$1 \times (c) = 0.0020$			$1 \times (c) = 0.0020$
FN2	0.015	0.9997	0.0150	28.013	0.4203	0.0246			$1 \times (c) = 0.0150$	
FCH4	0.994	0.9981	0.9458	16.043	15.1734	0.8869	$1 \times (c) = 0.9458$	$2 \times (c) = 1.8916$		
FC2H6	0.030	0.9916	0.0303	30.070	0.9100	0.0532	$2 \times (c) = 0.0606$	$3 \times (c) = 0.0909$		
FC3H8	0.005	0.9820	0.0051	44.097	0.2245	0.0131	$3 \times (c) = 0.0153$	$4 \times (c) = 0.0204$		
FC4H10	0.002	0.9661	0.0021	58.124	0.1203	0.0070	$4 \times (c) = 0.0084$	$5 \times (c) = 0.0105$		
FC5H12	0.001	0.9435	0.0011	72.151	0.0765	0.0045	$5 \times (c) = 0.0055$	$6 \times (c) = 0.0066$		
FC6H14	0.001	0.9120	0.0011	86.179	0.0945	0.0055	$6 \times (c) = 0.0066$	$7 \times (c) = 0.0077$		
Column totals:					17.1080	1.0000	1.0442	2.0277	0.0150	0.0020
Proportions by mass of elements in fuel (kg/kg):							$\dfrac{\Sigma(g) \times 12.011}{\Sigma(e)}$	$\dfrac{\Sigma(h) \times 2.016}{\Sigma(e)}$	$\dfrac{\Sigma(i) \times 28.013}{\Sigma(e)}$	$\dfrac{\Sigma(j) \times 31.998}{\Sigma(e)}$
							= 0.733	= 0.239	= 0.024	= 0.004

Appendix 7. Calculation of combustion air requirements and theoretical products of combustion for fuels specified on a mass basis

The calculation of these data is of interest, not only from the aspect of plant design, but also from the standpoint of operation and testing because the data indicate target figures for VCO2 and VO2 which might be attained when full combustion of the fuel is achieved with the specified amount of excess air.

Table A7.1 illustrates the procedure for carrying out the calculation, the fuel chosen being the Rank 601 washed smalls coal in Table A3.1.

From the table, the stoichiometric mass of air required $= 0.3121 \times 28.964$
$= 9.04$ kg/kg fuel

where 28.964 is the equivalent relative molecular mass of air.

Actual mass of air supplied $= 0.4370 \times 28.964$
$= 12.66$ kg/kg fuel

If the air volume is required, the calculation is based upon the fact that 1 kg mol of any gas occupies 22.414 m^3 at atmospheric pressure and 0°C. Using the gas laws to correct for an ambient temperature T1 of 20°C, 1 kg mol occupies 24.06 m^3.

Stoichiometric volume of air required $= 0.3121 \times 24.06 = 7.51$ m^3/kg of fuel.

Actual volume of air supplied $= 0.4370 \times 24.06 = 10.51$ m^3/kg of fuel. From the table, the flue gas constituents comprise:

Total wet products $= 0.0269$ kg mol/kg of fuel.
Total dry products $= 0.0564 + 0.3457 + 0.0262 + 0.0005$
$= 0.4288$ kg mol/kg fuel

Dry flue-gas analysis:

$$VCO2 = \frac{0.0564}{0.4288} \times 100 = 13.15\%$$

$$VN2 = \frac{0.3457}{0.4288} \times 100 = 80.62\%$$

Table A7.1 Calculation of the combustion air requirements and theoretical products of combustion for fuels specified on a mass basis

| | Fuel constituent | | | | | | Products of combustion kgmol/kg of fuel | | | | |
Symbol	Proportion kg/kg fuel (a)	Relative molecular mass (b)	kgmol/kg fuel (c)=(a)/(b)	Oxygen required kgmol/kgmol constituent (d)	Oxygen required kgmol/kg fuel (e)=(c)×(d)	Air required kgmol/kg fuel	CO2 1×(c)	N2 1×(c)	O2 1×(c)	SO2 1×(c)	H2O 1×(c)
MC	0.678	12.011	0.0564	1	0.0564		0.0564				
MH2	0.042	2.016	0.0208	0.5	0.0104						0.0208
MH2O	0.110	18.015	0.0061								0.0061
MN2	0.014	28.013	0.0005					0.0005			
MS	0.017	32.066	0.0005	1	0.0005					0.0005	
MA	0.080										
MO2	0.059	31.998	0.0018		−0.0018						

Stoichiometric oxygen = Σ(e): 0.0656

Stoichiometric air = Σ(e) × 100/21: 0.3121

Nitrogen in stoichiometric air = Σ(e) × 79/21: 0.2466

Stoichiometric air and products of combustion: 0.3121 0.0564 0.2471 0.0005 0.0269

If, say, 40% excess air is supplied, this will add: 0.1249 0.0986 0.0262

Final theoretical air and products of combustion: 0.4370 0.0564 0.3457 0.0262 0.0005 0.0269

$$VO2 = \frac{0.0262}{0.4288} \times 100 = 6.11\%$$

$$VSO2 = \frac{0.0005}{0.4288} \times 100 = \underline{0.12\%}$$
$$100.00\%$$

Appendix 8.   Calculation of combustion air requirements and theoretical products of combustion for fuels specified on a volumetric basis

This procedure is included for interest, although as explained in Appendix 6, conversion of gaseous fuel specifications to mass bases is convenient when it is desired to use the same test formulae for solid, liquid and gaseous fuels.

Table A8.1 illustrates the procedure for carrying out the calculation, the fuel chosen being the North Sea natural gas in Table A3.1.

The densities of gaseous fuels are declared at standard atmospheric pressure and 15°C and for the North Sea gas under consideration is given as 0.723 kg/m^3. The density of air at 0°C is 1.293 kg/m^3, and may be taken as 1.226 kg/m^3 at 15°C.

Thus, as it is known from the table that 1 m^3 of fuel requires 11.218 m^3 of air at the same conditions of pressure and temperature, the equivalent ratio on a mass basis is that 0.723 kg of fuel requires 11.218×1.226 kg air.

$$\text{That is, 1 kg of fuel requires } \frac{11.218 \times 1.226}{0.723}$$

$$= 19.02 \text{ kg air}$$

From the table, flue-gas constituents comprise:

Total wet products $= 2.021$ m^3/m^3 fuel

Total dry products $= 1.040 + 8.877 + 0.307 = 10.224$ m^3/m^3 fuel

Dry flue-gas analysis:

$$VCO2 = \frac{1.040}{10.224} \times 100 = \quad 10.17\%$$

$$VN2 \quad = \frac{8.877}{10.224} \times 100 = \quad 86.83\%$$

$$VO2 \quad = \frac{0.307}{10.224} \times 100 = \quad \underline{3.00\%}$$

$$100.00\%$$

Table A8.1 Calculation of the combustion air requirements and theoretical products of combustion for fuels specified on a volumetric basis

Fuel constituent Symbol	Proportion m³/m³ fuel (a)	Oxygen required m³/m³ constituent (b)	Oxygen required m³/m³ fuel (c) = (a) × (b)	Air required m³/m³ fuel	Products of combustion m³/m³ fuel CO2	N2	O2	H2O
FCO2	0.002				1 × (a) = 0.002			
FN2	0.015					0.015		
FCH4	0.944	2	1.888		1 × (a) = 0.944			2 × (a) = 1.888
FC2H6	0.030	3.5	0.105		2 × (a) = 0.060			3 × (a) = 0.090
FC3H8	0.005	5	0.025		3 × (a) = 0.015			4 × (a) = 0.020
FC4H10	0.002	6.5	0.013		4 × (a) = 0.008			5 × (a) = 0.010
FC5H12	0.001	8	0.008		5 × (a) = 0.005			6 × (a) = 0.006
FC6H14	0.001	9.5	0.010		6 × (a) = 0.006			7 × (a) = 0.007
Stoichiometric oxygen = Σ(c):			2.044					
Stoichiometric air = Σ(c) × 100/21:				9.755				
Nitrogen in stoichiometric air = Σ(c) × 79/21:						7.706		
Stoichiometric air and products of combustion:				9.755	1.040	7.721		2.021
If, say, 15% excess air is supplied, this will add:				1.463		1.156	0.307	
Final theoretical air and products of combustion:				11.218	1.040	8.877	0.307	2.021

Appendix 9. Calculation of combustion air requirements and theoretical products of combustion for fuels specified on a mass basis and with allowance made for arbitrarily selected quantities of unburnt fuel

This calculation follows the same procedure as in Appendix 7 but arbitrarily selected quantities of carbon in the fuel have been assumed to remain unburnt as carbon in solid residues or to be burnt only partially to carbon monoxide. This has been done to obtain the flue-gas "test data" included in Table A3.2 and used in the worked examples in the text and other Appendices. (It should be noted that the proportion of carbon assumed here to be burnt to carbon monoxide would be considered excessive when operating the more efficient types of combustion appliance with effective controls).

Table A9.1 illustrates the procedure for carrying out the calculation, the fuel chosen being the Rank 601 washed smalls coal in Table A3.1.

It should be noted that an alternative procedure could have been followed, with the fuel analysis adjusted to allow for the reduction in carbon content due to the carbon remaining unburnt in solid residues. This method would have yielded the same results, but it is more straightforward to base combustion calculations on the "as-fired" analysis of the fuel.

Note also that the "stoichiometric" oxygen supplied is that obtained as shown in Table A7.1, in which full combustion is assumed. Comparing Tables A7.1 and A9.1 then shows the effect that incomplete combustion has on the products of combustion for the same fuel/air ratio.

From the table, air supplied = 0.3121×28.964
$$= 9.04 \text{ kg/kg fuel}$$

where 28.964 is the equivalent relative molecular mass of air.

Total air supplied = 0.4370×28.964
$$= 12.66 \text{ kg/kg fuel}$$

From the table, the flue gas constituents comprise:

Total wet products = $0.0269 \text{ kg mol/kg fuel}$.

Total dry products = $0.0524 + 0.0020 + 0.3457 + 0.0293 + 0.0005$
$$= 0.4299 \text{ kg mol/kg fuel}$$

Dry flue-gas analysis:

$$VCO2 = \frac{0.0524}{0.4299} \times 100 = 12.19\%$$

$$VCO = \frac{0.0020}{0.4299} \times 100 = 0.46\%$$

$$VN2 = \frac{0.3457}{0.4299} \times 100 = 80.41\%$$

$$VO2 = \frac{0.0293}{0.4299} \times 100 = 6.82\%$$

$$VSO2 = \frac{0.0005}{0.4299} \times 100 = \underline{0.12\%}$$
$$100.00\%$$

Using this flue-gas analysis, and the known relative molecular masses of the constituents, the equivalent relative molecular mass of the dry flue gas

$$= (12.19 \times 44.010) + (0.46 \times 28.011) + (80.41 \times 28.013) +$$
$$+ (6.82 \times 31.998) + (0.12 \times 64.070)/100$$
$$= 30.278$$

Table A9.1 Calculation of the combustion air requirements and theoretical products of combustion for fuels specified on a mass basis and with allowance made for arbitrarily selected quantities of unburnt fuel

	Fuel constituent						Products of combustion kg mol/kg of fuel					
Symbol	Proportion kg/kg fuel (a)	Relative molecular mass (b)	kg mol/kg fuel (c) = (a)/(b)	Oxygen required kg mol/kg mol constituent (d)	Oxygen required kg mol/kg fuel (e) = (c)×(d)	Air required kg mol/kg fuel	CO_2 1×(c)	CO 1×(c)	N_2 1×(c)	O_2 1×(c)	SO_2 1×(c)	H_2O 1×(c)
MC	0.678	12.011	0.0020 unburnt									
			0.0524 to CO_2	1	0.0524		0.0524					
			0.0020 to CO	0.5	0.0010			0.0020				
MH2	0.042	2.016	0.0208	0.5	0.0104							0.0208
MH2O	0.110	18.015	0.0061									0.0061
MN2	0.014	28.013	0.0005						0.0005			
MS	0.017	32.066	0.0005	1	0.0005						0.0005	
MA	0.080											
MO2	0.059	31.998	0.0018	1	−0.0018							
Oxygen used in combustion = Σ(e):					0.0625							
Stoichiometric oxygen supplied (from Table A7.1):					0.0656							
Surplus oxygen from stoichiometric quantity:										0.0031		
Stoichiometric air (from Table A7.1):						0.3121						
Nitrogen in stoichiometric air (from Table A7.1):									0.2466			
Stoichiometric air and products of combustion:						0.3121	0.0524	0.0020	0.2471	0.0031	0.0005	0.0269
If, say, 40% excess air is supplied, this will add:						0.1249			0.0986	0.0262		
Final theoretical air and products of combustion:						0.4370	0.0524	0.0020	0.3457	0.0293	0.0005	0.0269

Appendix 10.   Calculation of mass of dry flue gas per unit mass of fuel

The quantity of dry flue gas, MG, is a factor in several formulae used in boiler test calculations. This, of course, is the quantity of flue gas actually produced when operating the boiler, not the theoretical quantity which can be calculated as in Appendices 7 and 8. The calculation is carried out by first selecting a substance in the fuel which can be quantified by analysis and then determining its proportion in the dry flue gas.

With this information, and using kg mol and kg units

$$MG = \frac{kg\,mol\ of\ substance\ in\ 1\ kg\,mol\ of\ dry\ flue\ gas}{kg\,mol\ of\ substance\ in\ 1\ kg\ of\ fuel}$$

Clearly, the substance selected should not also be present in the air supplied for combustion, so oxygen and nitrogen cannot be chosen. In solid and liquid fuels sulphur is usually present, but often in very small quantities and this would mean if it were to be selected for use in the above equation it would have to be quantified with great precision, both as a proportion of the mass of fuel, and as unburnt sulphur in the solid residues. Further difficulty would arise in attempting to measure the proportion of sulphur-bearing gases in the flue gas.

Carbon is the substance upon which the calculation is based and if all of the carbon in the fuel is burned, and all of the carbon-bearing gases in the flue gas accurately measured

$$MG = \frac{MC}{12.011} \times \frac{1}{kg\,mol\ of\ carbon\ in\ 1\ kg\,mol\ of\ flue\ gas}$$

In fact, in the case of solid and liquid fuels, not all of the carbon is burnt. Some of it, amounting in total to MCR kg/kg of fuel, is lost in the solid residues, thus reducing the carbon actually burnt to $(MC - MCR)$ kg/kg of fuel.

The treatment of the flue-gas analysis in order to obtain the carbon content of the dry flue gas is illustrated in Table A10.1, column (a) being the flue-gas analysis derived from Table A9.1.

Note that in practical tests VSO2 is unlikely to be measured. Assuming that a true measurement of VCO2 has been obtained, then VSO2 will appear as part of VN2, which is obtained by difference. This does not affect the accuracy with which the carbon content of the dry flue gas is obtained.

Table A10.2 repeats the procedure, but in this case the value of VCO2 is increased by adding on VSO2, which occurs when absorptive (Orsat) type instruments are used for flue-gas analysis.

It will be observed that for the products of combustion listed the tables are not really necessary because the kg mols of carbon are equivalent to the sum of the percentages of carbon-bearing gases, converted to simple proportions by dividing by 100. This follows from the structure of the molar combustion reactions

$$1 \text{ kg mol C} + 1 \text{ kg mol O2} \longrightarrow 1 \text{ kg mol CO2}$$

$$\text{and} \quad 1 \text{ kg mol C} + \tfrac{1}{2} \text{ kg mol O2} \longrightarrow 1 \text{ kg mol CO}$$

Therefore the carbon content of the flue gas is given by

$$\frac{\text{VCO2} + \text{VCO}}{100}$$

From the foregoing, therefore

$$\text{MG} = \frac{\text{MC} - \text{MCR}}{12.011} \times \frac{100}{\text{VCO2} + \text{VCO}}$$

Table A10.1 Calculation of the carbon quantity in dry flue gas using a true value for VCO2

Flue-gas constituent	Proportion in dry flue gas % (a)	Carbon content of dry flue gas kg mol/kg mol dry flue gas (b) = (a)/100
VCO2	12.19	0.1219
VCO	0.46	0.0046
VN2	80.41	
VO2	6.82	
VSO2	0.12	
	Σ(a) = 100.00	Σ(b) = 0.1265

Table A10.2 Calculation of the carbon quantity in the dry flue gas using a value for VCO2 which includes VSO2

Flue-gas constituent	Proportion in dry flue gas % (a)	Carbon content of dry flue gas kg mol/kg mol dry flue gas (b) = (a)/100
VCO2	12.31	0.1231
VCO	0.46	0.0046
VN2	80.41	
VO2	6.82	
	Σ(a) = 100.00	Σ(b) = 0.1277

In simpler test codes (for example, Refs. 1, 2 and 3) VCO2 and VCO are the only carbon-bearing gases in the flue gas which are considered. More detailed tests codes (for example, Refs. 4 and 5) allow for the possible presence of hydrocarbon gases designated by the general formula CXHY. Each value of VCXHY is added to the factor (VCO2 + VCO) which is then shown as (VCO2 + VCO + ΣVCXHY).

Again, in tests where absorptive-type instruments are used to measure VCO2, yielding results represented by Table A10.2, a correction may be applied to compensate for the inclusion of VSO2 with VCO2, as follows.

Note that, as remarked previously, MG can also be derived from the sulphur in the fuel and the sulphur dioxide in the flue gas. This can be used to relate carbon and sulphur quantities.

Assuming full combustion of both carbon and sulphur

$$MG = \frac{MS \times 100}{32.066 \times VSO2} = \frac{MC \times 100}{12.011 \times VCO2}$$

$$\text{Thus } VSO2 = \frac{12.011}{32.066} \times \frac{MS}{MC} \times VCO2 = \frac{VCO2}{2.67} \times \frac{MS}{MC}$$

$$\text{and } VCO2 + VSO2 = VCO2 + \left(\frac{VCO2}{2.67} \times \frac{MS}{MC} \right)$$

$$= VCO2 \left(1 + \frac{MS}{2.67\,MC} \right)$$

This shows that the effect of adding VSO2 to VCO2 is equivalent to multiplying VCO2 by

$$1 + \frac{MS}{2.67\,MC}$$

Therefore, to obtain the true VCO2 when VCO2 actually comprises (VCO2 + VSO2), the latter should be divided by

$$1 + \frac{MS}{2.67\,MC}$$

$$\text{Thus } MG = \frac{100MC \left(1 + \frac{MS}{2.67} \right)}{12.011 \times VCO2}$$

$$= \frac{100 \left(MC + \frac{MS}{2.67} \right)}{12.011 \times VCO2}$$

Expanding this to allow for unburnt carbon, the presence of carbon monoxide, and (if required) the presence of other unburnt hydrocarbon gases designated by $\Sigma CXHY$, gives the general expression

$$MG = \frac{100 \times \left(MC - MCR + \dfrac{MS}{2.67}\right)}{12.011 \times (VCO2 + VCO + \Sigma VCXHY)}$$

The sulphur, like the carbon, is never completely burnt in practice, but the correction for sulphur derived on this basis is sufficiently accurate for boiler test calculations, especially when it is borne in mind that the proportion of sulphur in most fuels is often relatively small.

Using the data given in Tables A10.1 and A10.2, in which full combustion of sulphur is assumed, the effect of using the formula for MG in its various forms can be illustrated.

(a) When a true reading of VCO2 is provided by the flue-gas analysis equipment

$$MG = \frac{100 \times (MC - MCR)}{12.011 \times (VCO2 + VCO)}$$

$$= \frac{100 \times (0.678 - 0.0244)}{12.011 \times (12.19 + 0.46)}$$

$$= 0.430 \text{ kg mol/kg fuel}$$

This compares well with the value 0.4299 derived in Appendix 9.

(b) When the reading of VCO2 is actually $(VCO2 + VSO2)$ but no correction is made for this.

$$MG = \frac{100 \times MC}{12.011 \times (VCO2 + VCO)}$$

$$= \frac{100 \times 0.678}{12.011 \times (12.31 + 0.46)}$$

$$= 0.426 \text{ kg mol/kg fuel}$$

(c) When the reading of VCO2 is actually $(VCO2 + VSO2)$ and a correction is made for this.

$$MG = \frac{100 \times \left(MC - MCR + \dfrac{MS}{2.67}\right)}{12.011 \times (VCO2 + VCO)}$$

$$= \frac{100 \times \left(0.678 - 0.0244 + \dfrac{0.017}{2.67}\right)}{12.011 \times (12.31 + 0.46)}$$

$$= 0.430 \; \text{kg mol/kg fuel}$$

The values calculated for MG in (a), (b) and (c) show that whilst the correction required for sulphur when using absorptive flue-gas analysis equipment may be small, it can be effectively achieved using the method in (c).

Appendix 11. Calculation of mean specific heat of dry flue gas

In boiler test codes, use of an average value for the mean specific heat S41 of the dry flue gas may be recommended, or methods of calculation, graphs, or nomograms may be provided to enable it to be determined. Sometimes, as in the case of gaseous fuels comprising constituent gases mixed in known proportions, the specific heat of the entire theoretical products of combustion, including moisture, may be declared. However, it is usual in boiler test work for the "dry" and "wet" products to be treated separately because it is informative to quantify the losses due to each, and therefore it is the specific heat of the dry flue gas which is required.

Table A11.1 illustrates the procedure for calculating the mean specific heats of dry flue gas over the temperature ranges 0–T1°C and 0–T2°C. The example chosen is based upon the flue-gas analysis derived from Table A9.1, together with the data provided in Appendix 3.

Having calculated the mean specific heat of the dry flue gas for the temperature ranges 0–20°C and 0–150°C, the mean specific heat, S41 for the temperature range 20–150°C

Table A11.1 Calculation of the mean specific heat of dry flue gas

Flue-gas constituent	Proportion dry flue gas % (a)	Mean sp. heat 0–20°C* kJ/kg mol K (b)	Contribution to mean sp. heat flue gas 0–20°C kJ/kg mol K $(c) = \dfrac{(a) \times (b)}{100}$	Mean sp. heat 0–150°C* kJ/kg mol K (d)	Contribution to mean sp. heat flue gas 0–150°C kJ/kg mol K $(e) = \dfrac{(a) \times (d)}{100}$
VCO2	12.19	36.598	4.461	39.413	4.804
VCO	0.46	29.184	0.134	29.231	0.134
VN2	80.41	29.184	23.467	29.231	23.505
VO2	6.82	29.372	2.003	29.747	2.029
VSO2	0.12	43.669	0.052	45.053	0.054
		$\Sigma(c) =$	30.073	$\Sigma(e) =$	30.526

*Interpolated principally from data in Ref. 10.

102

$$= \frac{(150 \times 30.5262) - (20 \times 30.0731)}{150 - 20}$$

$$= 30.6 \text{ kJ/(kg mol K)}$$

Reference to published values of gaseous specific heats such as those in Ref. 10 shows that in the case of nitrogen, oxygen and carbon monoxide, there is little variation in the temperature range 0–200°C. For carbon dioxide, the value rises more steeply, but even allowing for this, it is not unreasonable for a single, standard, specific heat to be taken for the flue gases of boilers operating with normal fuels in this low gas temperature range.

In Refs. 4 and 5 a value of 30.6 kJ/(kg mol K) is recommended for S41. Ref. 7 provides graphs from which a mean specific heat can be derived from a knowledge of the carbon/hydrogen ratio of the fuel and the flue-gas temperature. The specific heat of the gas does not, of course, appear directly in test codes such as Refs. 1, 2 and 3, where it is hidden in the overall constant K41 used in the formula for L41, but it is possible to estimate the value which has been assumed from the values given for K41. This has been done to obtain the values for S41 listed in Table A3.3. To make comparison easier, conversion from the original units has been carried out using the following conversion factors, as appropriate.

$$1 \text{ kJ/(kg mol K)} = \text{kcal/(kg K)} \times 4.1868 \times \text{RMM}$$

$$= \text{kcal/(m}^3\text{K)} \times 4.1868 \times 22.414$$

$$= \text{Btu/(lb R)} \times 4.1868 \times \text{RMM}$$

$$= \text{Btu/(lb mol R)} \times 4.1868$$

$$= \text{kJ/(kg K)} \times \text{RMM}$$

where RMM is the relative molecular mass of the flue gas. The flue gas in Appendix 9 has a relative molecular mass of 30.278.

Appendix 12. Calculation of loss due to sensible heat in dry flue gas

Given the quantity of dry flue-gas MG (see Appendix 10) and its mean specific heat S41 (see Appendix 11), datum temperature T1, and the flue-gas exit temperature T2, the loss due to the sensible heat in the dry flue gas is given by

$$L41 = \frac{MG \times S41 \times (T2 - T1) \times B11 \times 100}{H1}$$

or, expanding MG,

$$L41 = \frac{100 \times \left(MC - MCR + \dfrac{MS}{2.67} \right) \times S41 \times (T2 - T1) \times B11 \times 100}{12.011 \times (VCO2 + VCO + \Sigma VCXHY) \times H1}$$

This is a fundamental formula, which, in any specific case, may involve fewer factors than those included above. For example, for fuels other than solid fuel, MCR may be zero or negligibly small; the flue-gas analysis may ignore $\Sigma VCXHY$ if the quantity of hydrocarbon gases is also considered negligible; MS/2.67 will not be included unless the flue-gas analysis has been determined using absorptive-type equipment which gives a combined reading of VCO2 with VSO2. Furthermore, the heat input H1 may comprise only that heat resulting from the calorific value of the fuel.

In 1888, A. Siegert, of Munich (Ref. 15), noted that investigations into the properties of coals and their combustion showed that, when combustion was complete,

(i) the calorific value of the fuel divided by the carbon content was nearly constant,
(ii) the stoichiometric quantity of air required for combustion divided by the carbon content of the fuel was nearly constant,
(iii) the percentage of carbon dioxide in the flue gas, multiplied by the ratio of actual to stoichiometric air, was nearly constant.

Siegert took the mean specific heat of the flue gas as being constant, which, as has been noted in Appendix 11, is a reasonable assumption for the range of constituents and temperatures common in boiler practice. He was thus able to derive an extremely simple formula for the loss due to sensible heat in the flue gas, as follows:

$$L41 = \frac{K41 \times (T2 - T1)}{VCO2}$$

where K41 is an overall constant which combines the individual constants.

If this seems remarkable, consider again the fundamental expression for L41, but simplified as a result of assuming full combustion, ignoring sulphur, and assuming that H1 comprises no more than that due to calorific value, that is, $(Q11 \times B11)$.

$$L41 = \frac{(100)^*}{(12.011)^*} \times \left(\frac{MC}{Q11}\right)^* \times \frac{(S41)^*}{(VCO2)} (T2 - T1) \times \frac{(B11)^*}{(B11)^*} \times (100)^*$$

It will be noted that all factors marked with an asterisk are, following the Siegert approach, either constant or nearly constant. The expression therefore reduces to

$$L41 = \frac{K41 \times (T2 - T1)}{VCO2}$$

Siegert's investigations were concerned only with coals but the convenience of his method led to his formula being adapted for use with other fuels, the constant being adjusted to suit the fuel.

If the shortened Siegert version of L41 is equated to the full version, as follows:

$$\frac{K41 \times (T2 - T1)}{VCO2} = \frac{100}{12.011} \times \frac{MC}{Q11} \times \frac{S41}{VCO2} \times (T2 - T1) \times \frac{B11}{B11} \times 100$$

it then follows that

$$K41 = \frac{100 \times MC \times S41 \times 100}{12.011 \times Q11}$$

In Refs. 4 and 5, a mean value for S41 of 30.6 kJ/kg mol K is used. If the same value is used in the Siegert formula, values for the constant K41 can be derived from

$$K41 = \frac{100 \times 100 \times 30.6 \times MC}{12.011 \times Q11}$$

$$= 25\,477 \times \frac{MC}{Q11}$$

This applies to any fuel, provided the carbon content and the calorific value are known as a mass basis. For gaseous fuels, normally specified on a volumetric basis, conversion of principal characteristics to a mass basis can easily be carried out as shown in Appendix 6. In test codes employing

Siegert formulae, such as Ref. 3, or its predecessors, Refs. 1 and 2, constants are provided for the principal fuels, and for gross and net calorific values.

For greater accuracy, the Siegert formula was subsequently modified to allow for incomplete combustion. The variation in MG due to the presence of carbon monoxide was still ignored, but unburnt carbon in solid residues was taken into account. Reference to the fundamental formula shows that to allow for unburnt carbon, MCR is subracted from MC, and this may be done very simply by modifying the constant K41 thus

$$\text{modified K41} = \frac{25\,477 \times (\text{MC} - \text{MCR})}{\text{Q11}}$$

If this procedure were followed, it would require the calculation of a new value of K41 every time L41 was calculated. Alternatively, the original value of K41 (appropriate to the given fuel) can be used in all cases, with a separate modifying factor introduced to allow for unburnt carbon. Thus, from the previous expression,

$$\text{modified constant} = \frac{25\,477 \times \text{MC}}{\text{Q11}} - \frac{25\,477 \times \text{MCR}}{\text{Q11}}$$

But this can be expressed in terms of the unmodified constant, that is

$$\text{modified constant} = \text{K41} \left(1 - \frac{\text{MCR}}{\text{MC}} \right)$$

In other words, the original constant K41 is still used but is multiplied by

$$\left(1 - \frac{\text{MCR}}{\text{MC}} \right)$$

However, in test codes which include the Siegert formula, the constant is actually modified by a factor which is based upon the percentage heat loss L67, due to unburnt carbon in solid residues. The reason for adopting this procedure is a little obscure but possibly stems from a reluctance to require test staff using standard Siegert constants to obtain a value for MC, although, as previously noted, this is useful anyway if an exact value for K41 is desired.

Another possible reason for using L67 is that the Siegert formula is very widely used for routine checks upon boiler performance, in which only the loss due to sensible heat in the dry flue gas (the largest single loss) is determined. If, from previous and more detailed tests the normal value for L67 is known, this can be assumed and inserted in the modified Siegert formula without having to give consideration to the carbon contents of the fuel and ash.

The correction factor based upon L67 is obtained in the following way.

Note that $L67 = \dfrac{MCR \times Q60 \times 100}{Q11}$

and MCR $= \dfrac{0.01 \times L67 \times Q11}{Q60}$

Substituting for MCR in the multiplying factor previously determined gives

$$\text{modified constant} = K41\left[1 - \left(\frac{0.01 \times L67 \times Q11}{MC \times Q60}\right)\right]$$

where $Q60 = 33\,820$ kJ/kg

Taking as an example the Rank 601 coal in Table A3.1,

$Q11 = 27\,800$ kJ/kg and $MC = 0.678$

$$\text{and the modified constant} = K41 \times \left[1 - \left(\frac{0.01 \times L67 \times 27\,800}{0.678 \times 33\,820}\right)\right]$$

$$= K41 \times [1 - (0.012\,L67)]$$

For that coal, therefore, the Siegert formula becomes

$$L41 = \frac{0.6213 \times (T2 - T1)[1 - (0.012\,L67)]}{VCO2}$$

If the factor by which K41 is modified is obtained for other coals, the multiplying factor of L67 remains reasonably steady at about 0.012 but is closer to 0.011 for high-rank coals. If solid residues resulting from the combustion of oil fuel are considered, the factor is 0.015.

In Ref. 1 the expression giving the equivalent of L41 shows the losses due to unburnt carbon multiplied by 0.012. Refs. 2, 3 and 6 have instead given 0.01. In all of these references, a table is provided, listing typical average values of K41, with coal being subdivided into "bituminous" and "anthracite". Ref. 3 also includes the method for deriving K41 from the fuel characteristics. In this test code the Siegert formula for bituminous coal appears as

$$L41 = \frac{0.62 \times (T2 - T1)[1 - (0.01\,L67)]}{VCO2}$$

Appendix 13. Calculation of loss due to unburnt gases in flue gas

(a) Carbon monoxide

The presence of carbon monoxide in the flue gas represents a heat loss because the carbon in the carbon monoxide should, under ideal circumstances, burn to carbon dioxide. The calculation of the heat loss involved may be carried out as follows:

Quantity of dry flue gas = MG (see Appendix 10)
Quantity of carbon monoxide

$$= \frac{VCO}{100} \times MG \text{ kg mol/kg fuel}$$

$$= \frac{28.011 \times VCO \times MG}{100} \text{ kg/kg fuel}$$

$$= \frac{28.011 \times VCO \times MG \times B11}{100} \text{ kg/s}$$

Heat loss in unburnt carbon monoxide

$$= \frac{28.011 \times VCO \times MG \times B11 \times Q42}{100} \text{ kJ/s}$$

Or, in relation to heat input, heat loss due to unburnt carbon monoxide

$$L42 = \frac{28.011 \times VCO \times MG \times B11 \times Q42 \times 100}{100 \times H1} \%$$

This can be presented with MG expanded to show its constituent factors, thus

$$L42 = \frac{28.011 \times VCO \times \left(MC - MCR + \dfrac{MS}{2.67}\right) \times B11 \times Q42 \times 100}{12.011 \times (VCO2 + VCO + \Sigma VCXHY) \times H1}$$

Similar expressions are found in Refs. 4 and 5.

Ref. 7 takes a different route in obtaining L42, as follows.

It was noted that the quantity of carbon monoxide in the flue gas

$$= \frac{VCO}{100} \times MG \quad \text{kg mol/kg fuel}$$

From the combustion reaction for carbon to carbon monoxide (Table A2.1) it is known that this quantity is also the quantity of carbon in the fuel which burns to carbon monoxide, and this, converted from kg mol to kg

$$= \frac{12.011 \times VCO \times MG}{100} \quad \text{kg/kg fuel}$$

$$= \frac{12.011 \times VCO \times MG \times B11}{100} \quad \text{kg/s}$$

Heat loss due to unburnt carbon monoxide (= heat loss in partially burnt carbon)

$$= \frac{12.011 \times VCO \times MG \times B11 \times (Q12 - Q13)}{100} \quad \text{kJ/s}$$

It should be noted that because of the different approach (deriving the heat loss from the quantity of partially burnt carbon rather than the quantity of carbon monoxide), the calorific value represented by (Q12-Q13) is used in place of Q42. Ref. 7 attributes a value of 10 160 Btu/lb (23 632 kJ/kg) to this quantity.

Thus, in relation to heat input, loss due to unburnt carbon monoxide

$$L42 = \frac{12.011 \times VCO \times MG \times B11 \times (Q12-Q13) \times 100}{100 \times H1} \%$$

Therefore, with MG expanded

$$L42 = \frac{12.011 \times VCO \times \left(MC - MCR + \dfrac{MS}{2.67} \right) \times \times B11 \times (Q12 - Q13) \times 100 \times 100}{100 \times 12.011 \times (VCO2 + VCO + \Sigma VCXHY) \times H1}$$

$$= \frac{VCO \times B11 \times (Q12 - Q13) \times 100}{(VCO2 + VCO + \Sigma VCXHY) \times H1}$$

Ref. 1 includes an abbreviated, Siegert-type formula for calculating L42. This is in the most basic form, that is, it assumes no unburnt carbon, thus

$$L42 = \frac{K42 \times VCO}{(VCO2 + VCO)}$$

A list of values for K42 for the principal fuels is provided, and 59 is suggested for bituminous coal, but as with the Siegert formula for L41 the accuracy of the simplified formula for L42 can be increased by calculating the value of K42 for each fuel under consideration, instead of taking the typical values.

If the simplified and fundamental versions of L42 are equated

$$\frac{K42 \times VCO}{(VCO2 + VCO)} = \frac{28.011 \times MC \times VCO \times B11 \times Q42 \times 100}{12.011 \times (VCO2 + VCO) \times B11 \times Q11}$$

And taking Q42 as 10 100 kJ/kg.

$$K42 = 2\ 355\ 433 \times \frac{MC}{Q11}$$

As an example, the Rank 601 coal in Table A3.1 would have the constant

$$K42 = 2\ 355\ 433 \times \frac{0.678}{27\ 800}$$

$$= 57.4$$

In Refs. 2, 3 and 6, allowance is made for unburnt carbon in solid residues, and again, the correction is based upon L67, thus

$$L42 = \frac{K42 \times VCO \times [1 - (0.01\ L67)]}{(VCO2 + VCO)}$$

As was shown in Appendix 12 for the Siegert formula for L41, slightly more accurate results would be obtained by replacing the factor 0.01 with the figures suggested there.

(b) Hydrocarbon gases

Combustible hydrocarbon gases designated by the general symbol CXHY may be found in boiler flue gases. Generally, it is considered that only traces may be found except when combustion is abnormally and obviously bad, that is, when smoke is being formed. Usually, in tests, the flue gas is not even monitored for CXHY gases but if their presence is noted and measured, each one is treated in similar fashion to carbon monoxide (4.4.3).

In place of the relative molecular mass of carbon monoxide (28.011) the appropriate value for the gas is used, this being $(12.011X + 1.008Y)$.

Similarly, the appropriate calorific value Q43 for the gas under consideration is substituted for Q42.

Thus, the heat loss due to unburnt methane, ethane or any other CHXY gas is

$$L43 = \frac{(12.011X + 1.008Y) \times VCXHY \times \times \left(MC - MCR + \dfrac{MS}{2.67} \right) \times B11 \times Q43 \times 100}{12.011 \times (VCO2 + VCO + \Sigma VCXHY)}$$

A similar formula is to be found in Refs. 4 and 5. Ref. 7 adopts a different approach. The mass of dry flue gas/lb fuel is divided by the specific mass of the flue gas (for which a formula is provided) to give the volume of flue gas/lb fuel. This is multiplied by the volume of hydrocarbon gas/ft^3 of flue gas, and by the calorific value of the hydrocarbon gas in Btu/ft^3. This results in the loss being declared in terms of Btu/lb of fuel, which is how all the other losses are declared. A similar formula is provided for unburnt hydrogen, which is given separate consideration.

Refs. 1, 2 and 3 do not consider any unburnt gases other than carbon monoxide.

Appendix 14. Calculation of the partial pressure of water vapour in flue gas

The partial pressure of the moisture, wet products, or water vapour in flue gas may be derived from the analysis of the fuel and flue gas, as follows.

$$\text{Quantity of dry flue gas} = MG$$

This can be calculated, as shown in Appendix 10, from the formula

$$MG = \frac{100 \times \left(MC - MCR + \dfrac{MS}{2.67}\right)}{12.011 \times (VCO2 + VCO + \Sigma VCXHY)}$$

Quantity of moisture in flue gas $= MW$

$$= \frac{8.936\,MH + MH2O}{18.015}$$

[Note that adjustment should be made for any other moisture injected into the furnace, or for any moisture removed in condensing units, if these quantities are significant in relation to $(MH2O + 8.936\,MH2)$.]

Therefore, with the total flue-gas pressure assumed to be atmospheric pressure P1 mbar absolute,

Partial pressure of moisture in the flue gas

$$P51 = \frac{MW \times P1}{(MG + MW)}$$

In cases where theoretical combustion calculations have been carried out and results obtained similar to those in Tables A7.1, A8.1 and A9.1, it is possible to calculate a theoretical value for P51.

Taking as an example the data calculated for the Rank 601 coal in Table 9.1, the pressure of the vapour in the flue gas is

$$P51 = \frac{0.0269 \times 1013.25}{(0.4299 + 0.0269)}$$

$$= 59.7 \text{ mbar}$$

In comparison, the value of P51 obtained when using the data calculated for the natural gas in Table A8.1 is

$$P51 = \frac{2.021 \times 1013.25}{(10.224 + 2.021)}$$

$$= 167.2 \text{ mbar}$$

If preferred, a value of P51 appropriate to the flue gas and fuel analysis may be calculated, as shown previously, for each test. Refs. 4 and 7 suggest taking 70 mbar absolute (1 lbf/in^2 absolute) as a reasonable average figure. As shown in 4.4.6, variations in the selected value for P51 make little difference when calculating L51.

Appendix 15.   Derivation of VCO2 from VO2

As shown in Appendix 10, it is necessary to know the proportion (or percentage) of carbon dioxide VCO2 in the dry flue gas in order to calculate the quantity of dry flue gas. In some tests VCO2 is directly measured. If, however, it is preferred to measure VO2, it then becomes necessary to derive VCO2 from this reading. The basis of the method is that, assuming the analysis of the fuel is known, the stoichiometric (zero excess air) VCO2 can be calculated by following the methods given in Appendices 7, 8 and 9, and the reduction in this figure due to the addition of excess air (as indicated by the VO2 reading) can be taken into account to give the actual VCO2.

Consider both stoichiometric and actual products of combustion in terms of, say kg mol/kg fuel, is shown in Fig. A15.1, with, for convenience, the symbols A, B and C representing the groups of products indicated.

$$\text{Stoichiometric VCO2} = \frac{A}{B} \times 100$$

Then assuming complete combustion of all carbon

$$\text{Actual VCO2} = \frac{A}{(B+C)} \times 100$$

The difference between stoichiometric and actual VCO2

$$= \left[\frac{A}{B} - \frac{A}{(B+C)} \right] \times 100$$

$$= \frac{A \times C \times 100}{B \times (B+C)}$$

Thus, for any level of excess air

$$\text{Actual VCO2} = \left[\frac{A}{B} - \frac{A}{B} \left(\frac{C}{B+C} \right) \right] \times 100$$

114

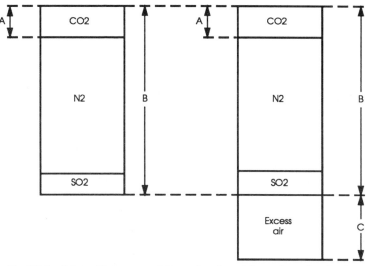

Fig. A15.1   Relationship between stoichiometric and actual dry products of combustion

But $\dfrac{C}{(B+C)}$ = proportion of excess air in flue gas

$$= \frac{VO2}{0.21 \times 100}$$

Therefore actual VCO2 = stoichiometric VCO2 × $\left(1 - \dfrac{VO2}{21}\right)$

This expression appears in Refs. 1, 2, 3 and 6, which include more simplified forms of test formulae. Ref. 3 includes a table of stoichiometric values of VCO2 for a number of fuels. Strictly, the stoichiometric value of VCO2 should be that pertaining to the fuel actually burnt. Allowance for unburnt combustible matter in solid residues can easily be made by subtracting MCR from MC before calculating the stoichiometric VCO2. It is more difficult in direct calculations to make allowance for carbon monoxide or other unburnt gas when measurable quantities of these are in the flue gas, but a computer can be programmed to carry out iterative calculations to obtain the quantity of carbon which is partially burned to, say, carbon monoxide in order to yield the measured values of VCO and VO2. As the same program could be arranged to give VCO2, use of the procedure given above would in those circumstances be unnecessary.

Appendix 16. Derivation of VCO2 and VO2 from VO2 (W)

Instruments are in use which give the proportion of oxygen in flue gas on the wet basis, that is, as the proportion of the combined dry gases and water vapour. This has to be converted to the dry basis for use in the formula used for calculating the quantity of dry flue gas.

As in Appendix 15, the calculation requires a knowledge of the theoretical products of combustion. Consider the total stoichiometric products of combustion (that is, including moisture) together with the excess nitrogen and oxygen in terms of kg mol/kg fuel. Note that the addition of (W) to any symbol indicates that the quantity is on the total or wet basis whilst (E) indicates that the quantity is supplied in excess air.

From Fig. A16.1

$$VO2(W) = \frac{O2(E)}{O2(E) + N2(E) + D}$$

Also, from the standard volumetric proportions of oxygen and nitrogen in air

$$N2(E) = O2(E) \times \frac{79}{21}$$

$$= 3.7619 \times O2(E)$$

$$\text{and } O2(E) + N2(E) = 4.762 \times O2(E)$$

$$\text{Thus } VO2(W) = \left[\frac{O2(E)}{4.762 \, O2(E) + D} \right] \times 100$$

$$\text{from which } O2(E) = \frac{D}{\left[\dfrac{100}{VO2(W)} - 4.762 \right]}$$

$$= \frac{(N2 + CO2 + SO2 + H2O)}{\left[\dfrac{100}{VO2(W)} - 4.762 \right]}$$

$$\text{and } N2(E) = O2(E) \times 3.7619$$

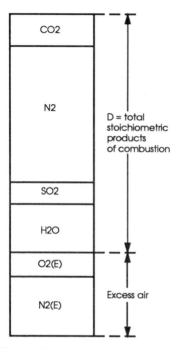

Fig. A16.1   Total stoichiometric products of combustion and excess air

Thus total actual dry products = MG

$$= \text{total stoichiometric dry products}$$

$$+ O2(E) + N2(E)$$

$$\text{Thus } VO2 = \frac{O2(E)}{MG} \times 100$$

$$\text{and } VCO2 = \frac{\text{stoichiometric CO2}}{MG} \times 100$$

These formulae do not appear in any of the references and have been included here for interest only. Strictly, the stoichiometric quantity of carbon dioxide used in the calculations should be that pertaining to the fuel actually burnt. Allowance can easily be made for unburnt combustible matter in solid residues by subtracting MCR from MC before carrying out the calculation of stoichiometric quantities. It is more difficult in direct calculations to make allowance for carbon monoxide or other unburnt gases when measurable quantities of these are in the flue gas, but

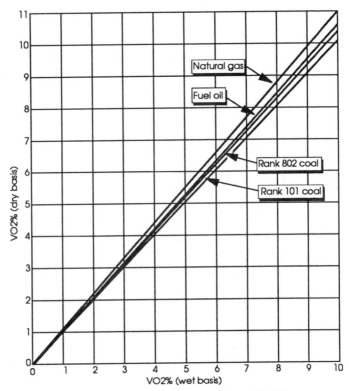

Fig. A16.2   Relationship between VO2 and VO2(W)

a computer can be programmed to carry out an iterative calculation to obtain the quantity of carbon which has partially burnt to, say, carbon monoxide in order to yield the measured values of VCO and VO2(W). As the same program could be arranged to give VO2 and VCO2, use of the procedure given above would in those circumstances be unnecessary.

Fig. A16.2 shows the relationship between VO2 and VO2(W) for a number of fuels, full combustion having been assumed.

Appendix 17. Calculation of excess air and total air supplied for combustion from test results and stoichiometric data

The basis of this procedure is that adjustment is made to the oxygen in the dry flue to allow for burning off any unburnt fuel and the excess air is calculated from the remaining quantity of excess oxygen. This is added to the known quantity of air required for stoichiometric combustion to give the total air supplied. The calculation may be carried out numerically, step by step, but a more general formula can be derived, as follows.

Oxygen required to burn off carbon monoxide

$$= \frac{VCO}{100} \times 0.5 \quad \text{kg mol/kg mol flue gas}$$

$$\text{Excess oxygen} = \frac{VO2}{100} - \left(\frac{VCO}{100} \times 0.5 \right) \quad \text{kg mol/kg mol flue gas}$$

Quantity of dry flue gas $=$ MG kg mol/kg fuel

Excess oxygen after carbon monoxide is burnt off

$$= MG \times \left(\frac{VO2}{100} - \frac{0.5 \, VCO}{100} \right) \quad \text{kg mol/kg fuel}$$

Oxygen required to burn off carbon in residues

$$= \frac{MCR}{12.011} \quad \text{kg mol/kg fuel}$$

Therefore excess oxygen after burning off both carbon and carbon monoxide

$$= \left[MG \times \left(\frac{VO2}{100} - \frac{0.5 \, VCO}{100} \right) \right] - \frac{MCR}{12.011} \quad \text{kg mol/kg fuel}$$

This expression is multiplied by the relative molecular mass of oxygen (31.998) to convert the units to kg oxygen/kg fuel, then divided by the

119

proportion of oxygen in the air (23.2% by mass) to obtain the mass of excess air, thus

$$\text{Excess air} = \left\{ \left[MG \times \left(\frac{VO2}{100} - \frac{0.5\,VCO}{100} \right) \right] - \frac{MCR}{12.011} \right\} \times \frac{31.998 \times 100}{23.2}$$

kg/kg fuel

Expanding MG (see Appendix 10) and simplifying

Excess air = 11.4337

$$\left\{ \left[\frac{\left(MC - MCR + \dfrac{MS}{2.67} \right)(VO2 - 0.5\,VCO)}{(VCO2 + VCO + \Sigma VCXHY)} \right] - MCR \right\}$$

kg/kg fuel

$MS/2.67$ is included only when absorptive-type flue-gas anlaysis equipment is used, giving a combined measurement of VCO2 and VSO2.

Illustrating the procedure with data drawn from the example in Appendix 3.

$$\text{Excess air} = 11.4337 \left[\frac{(0.678 - 0.0244)\left(6.82 - \dfrac{0.46}{2} \right)}{(12.19 + 0.46)} - 0.0244 \right]$$

= 3.61 kg/kg fuel

From table A7.1 it is known that the air required for stoichiometric combustion

= 9.04 kg/kg fuel

Therefore, total air supplied =   9.04 + 3.61

= 12.65 kg/kg fuel

This compares with 12.66 kg/kg fuel as calculated in Appendix 9.

Appendix 18. Calculation of total air supplied for combustion from VN2, MN2 and MG

When the total air supplied for combustion is required, this procedure may be considered simpler than that in Appendix 17 because it provides a direct method of calculation using test results alone, as follows:

$$\text{Quantity of dry flue gas} = MG \text{ kg mol/kg fuel}$$

$$\text{Nitrogen in flue gas} = \frac{VN2}{100} \times MG \text{ kg mol/kg fuel}$$

Nitrogen in flue gas from air supplied

$$= \left(\frac{VN2}{100} \times MG\right) - \frac{MN2}{28.013}$$

kg mol/kg fuel

Multiplying by the relative molecular mass of nitrogen (28.013) to convert the units to kg nitrogen/kg fuel, and dividing by the proportion of nitrogen in air (76.8% by mass) to obtain the mass of air,

Mass of air supplied

$$= \left(\frac{VN2 \times MG}{100} - \frac{MN2}{28.013}\right) \times \frac{28.013}{76.8} \text{ kg/kg fuel}$$

Expanding MG (see Appendix 10) and simplifying,

Air supplied = 1.3021

$$\left[\frac{2.332\, VN2 \times \left(MC - MCR + \dfrac{MS}{2.67}\right)}{(VCO2 + VCO + \Sigma VCXHY)} - MN2\right] \text{ kg/kg fuel}$$

MS/2.67 is included only when absorptive-type flue-gas analysis equipment is used, giving a combined measurement of VCO2 and VSO2.

Illustrating the procedure with data drawn from the example in Appendix 3,

$$\text{Air supplied} = 1.3021 \left[\frac{2.332 \times 80.41 \times (0.678 - 0.0244)}{(12.19 + 0.46)} - 0.014 \right]$$

$$= 12.6 \text{ kg/kg fuel}$$

This compares with 12.66 kg/kg fuel as calculated in Appendix 9.

Appendix 19. Calculation of mass flow rate of unweighed solid residue

In a test it may be inconvenient or impossible to weigh one of the types of solid residue being produced by the unit. If the other types of residue are collected and weighed, the quantity of the unweighed residue can be determined by difference, the total ash quantity being known from the firing rate and the ash content of the fuel burnt. In this procedure, it is necessary to distinguish between "ash" as it is collected from the unit, which always contains a proportion of combustible matter, and pure ash as specified in the ultimate analysis of the fuel, which is that proportion of ordinary ash which does not include any combustible matter. Thus, evaluating the production rate of pure ash

$$A61 = \frac{R61 \times (100 - C61)}{100}$$

$$A62 = \frac{R62 \times (100 - C62)}{100}$$

$$A63 = \frac{R63 \times (100 - C63)}{100}$$

$$A64 = \frac{R64 \times (100 - C64)}{100}$$

$$A65 = \frac{R65 \times (100 - C65)}{100}$$

A65 is included only if the mill(s) is included within the unit.
Therefore, balance of pure ash

$$A66 = (B11 \times MA) - (A61 + A62 + A63 + A64 + A65)$$

Note that in the case of on-site milling, with the mill outside the unit and with the fuel weighed prior to milling,

$$B11 = W11 - R65$$

123

Therefore equivalent production rate of unweighed residue

$$R66 = \frac{100 \times A66}{(100 - C66)}$$

Using the data provided for the example in Appendix 3, only ash and clinker R62, grit and dust R63, and unweighed residue R66 have to be considered. In this case the unweighed residue can be assumed to be fine dust lodged in various parts of the unit and passing up the chimney.

$$\text{Therefore } A62 = \frac{0.102\,34 \times (100 - 12.132)}{100}$$

$$= 0.0899 \text{ kg/s}$$

$$A63 = \frac{0.029 \times (100 - 53.52)}{100}$$

$$= 0.0135 \text{ kg/s}$$

Production rate of pure ash in unweighed residue

$$A66 = (1.33 \times 0.08) - (0.0899 + 0.013\,48)$$

$$= 0.0030 \text{ kg/s}$$

Therefore production rate of unweighed residue

$$R66 = \frac{100 \times 0.003\,02}{(100 - 60)}$$

$$= 0.007\,55 \text{ kg/s}$$

Appendix 20. Calculation of volume of flue gas

The volume of the flue gas from the unit is unlikely to be required in a test, but a method of estimating this quantity is presented here for interest. Both the dry gases and the water vapour are assumed to be combined and behave as an ideal gas. The volume is then calculated using the usual gas equation, thus:

$$VFG = \frac{K2 \times MGW \times (T2 + 273.15)}{P2}$$

where K2 is the univeral gas constant, 8.314 kJ/(kg mol K), and appears in the equation because the gas quantity MGW is in kg mol/kg fuel. P2 is assumed to be atmospheric pressure, 1013.25 mbar.

The following worked example uses the data from Appendix 3.

$$VFG = \frac{8.314 \times (0.4299 + 0.0269) \times (150 + 273.15) \times 10}{1013.25}$$

$$= 15.86 \text{ m}^3/\text{kg fuel}$$

Alternatively, in cases such as this, where the pressure is assumed to be approximately the same at both inlet to and discharge from the unit, it may be ignored altogether and a simple volume–temperature equation used. It is known that 1 kg mol of any gas at 0°C occupies 22.414 m³, so for 0.4568 kg mol

$$VFG = \frac{0.4568 \times 22.414 \times (150 + 273.15)}{273.15}$$

$$= 15.86 \text{ m}^3/\text{kg fuel}$$

Appendix 21.   Test accuracy

Some test codes (Refs. 3, 4 and 6) require that the thermal efficiency of a unit should be declared within specified limits, namely, ± 2 percentage points. This means that if the thermal efficiency is calculated to be 80%, the true figure is known to lie between 78 and 82%. Refs. 3 and 4 state that this requirement applies to "thermal performance tests" so presumably the output of the unit is expected to be declared to the same accuracy. Others (Refs. 5 and 7) specify required accuracies for inidividual test measurements but do not indicate an overall test accuracy and therefore it is presumably for the test personnel to calculate the accuracy of the result obtained, based upon the measuring accuracy of the instruments selected.

Strictly, the calculation of overall test accuracy should be carried out using statistical methods, in which the bias and error of each measurement is estimated and the importance or weight of each is established.

A simpler approach is often used in which the worst possible case is considered. In this admittedly unlikely situation, each individual measurement error is assumed to be in a direction which increases the overall error in either a positive or negative direction.

Consider, for example, a simple hot-water unit fired by coal. The thermal efficiency is given by the expression

$$Z = \frac{W21 \times (E26 - E21) \times 100}{B11 \times Q11}$$

Assume that E21 and E26 (the enthalpies of the water at entry and discharge from the unit respectively) are determined to accuracies which are in proportion to the accuracies of the respective temperatures T21 (say 150°C) and T26 (say 95°C).

If (A) denotes the percentage accuracy (or maximum possible error) for each measurement, then the following are the numerical values which might apply if for example, the instruments referred to are selected (see Table 5.3).

W21(A) = $\pm 0.25\%$ (turbine meter)

T21(A) = $\pm 0.1\%$ (resistance thermometer)

T26(A) = $\pm 0.1\%$ (resistance thermometer)

B11(A) = $\pm 0.2\%$ (batch weigher)

Q11(A) = $\pm 1.2\%$ (sampling of fuel + bomb calorimeter)

126

To simplify the calculation, it can be assumed that $T26 - T21$ is determined to an accuracy of $\pm 0.2\%$, that is, with twice the error expected of a single temperature measurement.

For highest efficiency, with all errors contributing to an increase in efficiency,

$$Z(\max) = \frac{W21\left[1 + \dfrac{W21(A)}{100}\right] \times (T26 - T21)\left[1 + \dfrac{(T26 - T21)(A)}{100}\right] \times 100}{\left[1 - \dfrac{B11(A)}{100}\right] \times Q11\left[1 - \dfrac{Q11(A)}{100}\right]}$$

$$= Z \times \frac{\left[1 + \dfrac{W21(A)}{100}\right] \times \left[1 + \dfrac{(T26 - T21)(A)}{100}\right]}{\left[1 - \dfrac{B11(A)}{100}\right] \times \left[1 - \dfrac{Q11(A)}{100}\right]}$$

Substituting the values given

$$Z(\max) = Z \times \frac{1.0025 \times 1.002}{0.998 \times 0.998}$$

$$= Z \times 1.018\,74$$

Thus at 80% efficiency, the true figure allowing for instrument error might actually be

$$Z(\max) = 80 \times 1.018\,74$$

$$= 81.45\%$$

Similarly, with all errors contributing to a decrease in efficiency, the true figure allowing for instrument error would be

$$Z(\min) = 80 \times \frac{0.9975 \times 0.998}{1.002 \times 1.012}$$

$$= 78.54\%$$

Thus, using the instrumentation specified, the thermal efficiency could be declared at $80\% + 1.45$
$\quad\quad\quad\quad\quad\quad\quad\quad\quad - 1.46$.

This type of calculation can be performed for any of the routes that might be taken in determining the output or thermal efficiency of a unit. It can be incorporated with ease into computer programs so that alternative test results are available, indicating the extreme values appropriate to the instrument accuracies entered as data input.

Appendix 22. Theoretical performance curves

Figures A22.1–A22.7 are included for interest. They show for a number of fuels the dry flue-gas losses L41 for excess air values ranging from 0 to 100% and for flue-gas temperatures T2 from 100 to 400°C in steps of 20°C (with inlet temperature T1 taken as 20°C). Complete combustion of the fuel has been assumed in each case.

For the same range of flue-gas temperatures, the losses L51 due to moisture in the fuel are shown in a separate list on each figure.

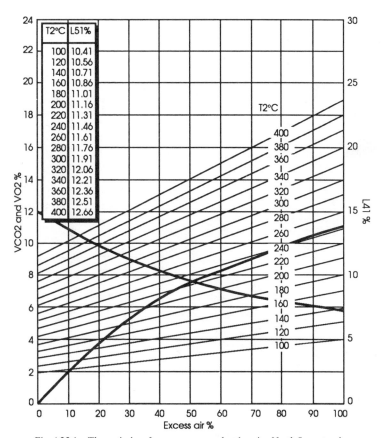

Fig. A22.1   Theoretical performance curves when burning North Sea natural gas

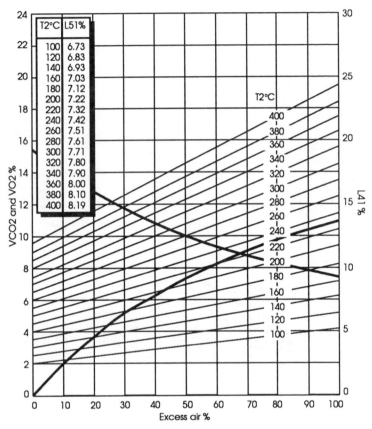

Fig. A22.2　Theoretical performance curves when burning Class D fuel oil

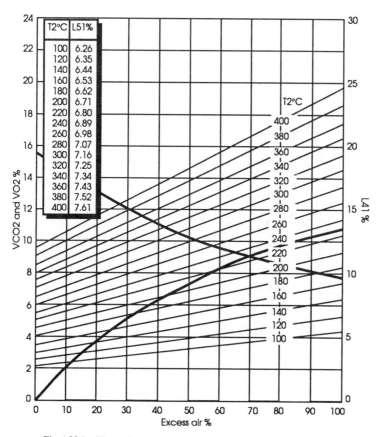

Fig. A22.3   Theoretical performance curves when burning Class E fuel oil

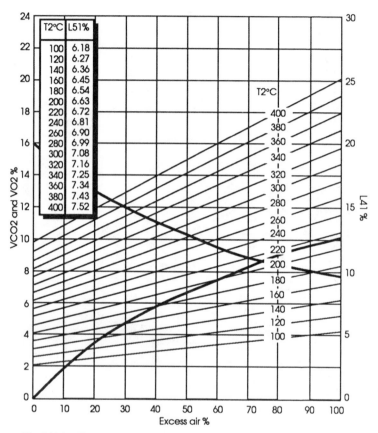

Fig. A22.4   Theoretical performance curves when burning Classes F and G fuel oils

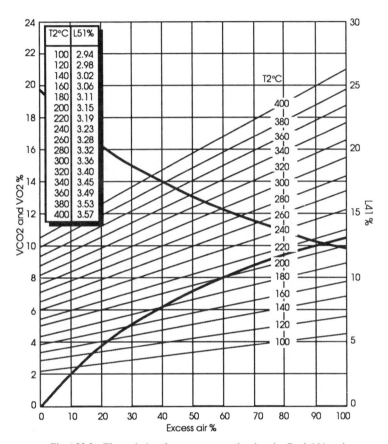

T2°C	L51%
100	2.94
120	2.98
140	3.02
160	3.06
180	3.11
200	3.15
220	3.19
240	3.23
260	3.28
280	3.32
300	3.36
320	3.40
340	3.45
360	3.49
380	3.53
400	3.57

Fig. A22.5   Theoretical performance curves when burning Rank 101 coal

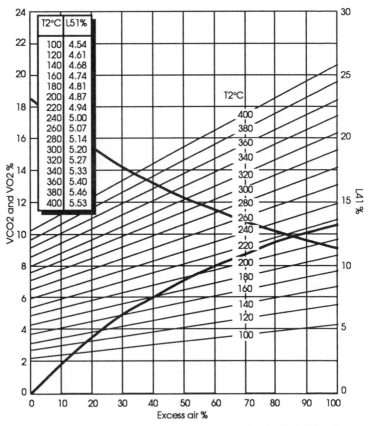

Fig. A22.6   Theoretical performance curves when burning Rank 601 coal

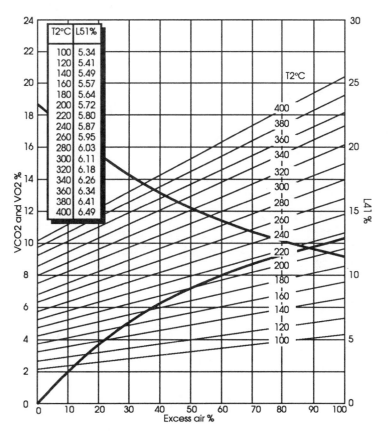

Fig. A22.7   Theoretical performance curves when burning Rank 802 coal

REFERENCES

References

1 BS 845: 1961. Code for acceptance tests for industrial type boilers and steam generators.*

2 BS 845: 1972. code for acceptance tests for industrial type boilers and steam generators.*

3 BS 845: 1987. Methods for assessing thermal performance of boilers for steam, hot water and high temperature heat transfer fluids. Part 1. Concise procedure.

4 BS 845: 1987. Methods for assessing thermal performance of boilers for steam, hot water and high temperature heat transfer fluids. Part 2. Comprehensive procedure.

5 BS 2885: 1974. Code for acceptance tests on stationary steam generators of the power station type.

6 BS DD65: Part 1. Direct method: 1979. Methods of type testing heating boilers for thermal performance. (At the time of writing, part of this document is being amended to form a new British Standard dealing with the testing of low-temperature hot-water boilers in ratings up to 600 kW.)

7 ASME PTC 4.1: 1985. Power test code — steam generating units.

8 BS 1469: 1962. Coal tar fuels.

9 BS 2869: 1983. Fuel oils for oil engines and burners for non-marine use.

10 *Technical Data on Fuel:* sixth edition: 1961. Edited by HM Spiers. British National Committee World Power Conference.

11 *Technical Data on Fuel:* seventh edition: 1977. Edited by J. W. Rose and J. R. Cooper. British National Committee World Energy Conference.

12 BS 1016 (21 parts, various dates). Methods for the analysis and testing of coal and coke.

13 BS 1017: Part 1: 1977. Sampling of coal.

*Ref. 1 was superseded by Ref. 2 which was superseded in turn by Refs. 3 and 4 but the earlier standards are retained as references to illustrate certain points in the historical background of Refs. 3 and 4.

14 *Technical Data on Solid Fuel Plant:* 1986. Published by British Coal. Produced by the College of Fuel Technology, London.

15 Determination of the heat loss in flue gases, by A. Siegert and W. Durr. *J. Gasbeleuchtung*, 1888, Vol. 12, p. 736.

16 BS 4947: 1984. Specification for test gases for gas appliances.

17 ASME PTC 3.2: 1985. Power test code — solid fuels.

18 ASME PTC 3.1: 1985. Power test code — diesel and burner fuels.